ENGINEERING FUNDAMENTALS

QUICK REFERENCE CARDS

Third Edition

Michael R. Lindeburg, P.E.

PROFESSIONAL PUBLICATIONS, INC.
Belmont, CA 94002

In the ENGINEERING REVIEW MANUAL SERIES

Engineer-In-Training Review Manual
  Engineering Fundamentals Quick Reference Cards
  Mini-Exams for the E-I-T Exam
  1001 Solved Engineering Fundamentals Problems
  E-I-T Review: A Study Guide
Civil Engineering Reference Manual
  Civil Engineering Quick Reference Cards
  Civil Engineering Sample Examination
  Civil Engineering Review Course on Cassettes
  Seismic Design for the Civil P.E. Exam
  Timber Design for the Civil P.E. Exam
Structural Engineering Practice Problem Manual
Mechanical Engineering Review Manual
  Mechanical Engineering Quick Reference Cards
  Mechanical Engineering Sample Examination
  101 Solved Mechanical Engineering Problems
  Mechanical Engineering Review Course on Cassettes
  Consolidated Gas Dynamics Tables
Electrical Engineering Reference Manual
Chemical Engineering Reference Manual
  Chemical Engineering Practice Exam Set
Land Surveyor Reference Manual
Metallurgical Engineering Practice Problem Manual
Petroleum Engineering Practice Problem Manual
Expanded Interest Tables
Engineering Law, Design Liability, and Professional Ethics
Engineering Unit Conversions

In the ENGINEERING CAREER ADVANCEMENT SERIES

How to Become a Professional Engineer
The Expert Witness Handbook—A Guide for Engineers
Getting Started as a Consulting Engineer
Intellectual Property Protection—A Guide for Engineers
E-I-T/P.E. Course Coordinator's Handbook
Becoming a Professional Engineer

Distributed by: Professional Publications, Inc.
1250 Fifth Avenue
Department 77
Belmont, CA 94002
(415) 593-9119

ENGINEERING FUNDAMENTALS QUICK REFERENCE CARDS
Third Edition

Printed in the United States of America

ISBN: 0-932276-88-1

Professional Publications, Inc.
1250 Fifth Avenue, Belmont, CA 94002

Current printing of this edition (last number): 6 5 4 3

TABLE OF CONTENTS

CONVERSION FACTORS

to convert	into	multiply by
acres	hectares	0.4047
acres	square feet	43,560.0
acres	square miles	1.562 EE–3
ampere hours	coulombs	3600.0
angstrom units	inches	3.937 EE–9
angstrom units	microns	1 EE–4
astronomical units	kilometers	1.495 EE8
atmospheres	cms of mercury	76.0
BTU's	horsepower-hrs	3.931 EE–4
BTU's	kilowatt-hrs	2.928 EE–4
BTU/hr	watts	0.2931
bushels	cubic inches	2150.4
calories, gram (mean)	BTU's (mean)	3.9685 EE–3
centares	square meters	1.0
centimeters	kilometers	1 EE–5
centimeters	meters	1 EE–2
centimeters	millimeters	10.0
centimeters	feet	3.281 EE–2
centimeters	inches	0.3937
chains	inches	792.0
coulombs	faradays	1.036 EE–5
cubic centimeters	cubic inches	0.06102
cubic centimeters	pints (U.S. liq.)	2.113 EE–3
cubic feet	cubic meters	0.02832
cubic feet/min	pounds water/min	62.43
cubic feet/sec	gallons/min	448.831
cubits	inches	18.0
days	seconds	86,400.0
degrees (angle)	radians	1.745 EE–2
degrees/sec	revolutions/min	0.1667
dynes	grams	1.020 EE–3
dynes	joules/meter (newtons)	1 EE–5
ells	inches	45.0
ergs	BTU's	9.480 EE–11
ergs	foot-pounds	7.3670 EE–8
ergs	kilowatt-hours	2.778 EE–14
faradays/sec	amperes (absolute)	96,500
fathoms	feet	6.0
feet	centimeters	30.48
feet	meters	0.3048
feet	miles (nautical)	1.645 EE–4
feet	miles (statute)	1.894 EE–4
feet/min	centimeters/sec	0.5080
feet/sec	knots	0.5921
feet/sec	miles/hour	0.6818
foot-pounds	BTU's	1.286 EE–3
foot-pounds	kilowatt-hours	3.766 EE–7
furlongs	miles (U.S.)	0.125
furlongs	feet	660.0
gallons	liters	3.785
gallons of water	pounds of water	8.3453
gallons/min	cubic feet/hour	8.0208
grams	ounces (avoirdupois)	3.527 EE–2
grams	ounces (troy)	3.215 EE–2
grams	pounds	2.205 EE–3
hectares	acres	2.471
hectares	square feet	1.076 EE5
horsepower	BTU's/min	42.44
horsepower	kilowatts	0.7457
horsepower	watts	745.7
hours	days	4.167 EE–2
hours	weeks	5.952 EE–3
inches	centimeters	2.540
inches	miles	1.578 EE–5
joules	BTU's	9.480 EE–4
joules	ergs	1 EE7
kilograms	pounds	2.205
kilometers	feet	3281.0
kilometers	meters	1,000.0
kilometers	miles	0.6214
kilometers/hr	knots	0.5396
kilowatts	horsepower	1.341
kilowatt-hours	BTU's	3413.0
knots	feet/hour	6080.0
knots	nautical miles/hr	1.0
knots	statute miles/hr	1.151
light years	miles	5.9 EE12
links (surveyor's)	inches	7.92
liters	cubic centimeters	1000.0
liters	cubic inches	61.02
liters	gallons (U.S. liq.)	0.2642
liters	milliliters	1000.0
liters	pints (U.S. liq)	2.113
meters	centimeters	100.0
meters	feet	3.281
meters	kilometers	1 EE–3
meters	miles (nautical)	5.396 EE–4
meters	miles (statute)	6.214 EE–4
meters	millimeters	1000.0
microns	meters	1 EE–6
miles (nautical)	feet	6080.27
miles (statute)	feet	5280.0
miles (nautical)	kilometers	1.853
miles (statute)	kilometers	1.609
miles (nautical)	miles (statute)	1.1516
miles (statute)	miles (nautical)	0.8684
miles/hour	feet/min	88.0
milligram/liter	parts/million	1.0
milliliters	liters	1 EE–3
millimeters	inches	3.937 EE–2
newtons	dynes	1 EE5
ohms (international)	ohms (absolute)	1.0005
ounces	grams	28.349527
ounces	pounds	6.25 EE–2
ounces (troy)	ounces (avoirdupois)	1.09714
parsecs	miles	19 EE12
parsecs	kilometers	3.084 EE13
pints (liq.)	cubic centimeters	473.2
pints (liq.)	cubic inches	28.87
pints (liq.)	gallons	0.125
pints (liq.)	quarts (liq.)	0.5
pounds	kilograms	0.4536
pounds	ounces	16.0
pounds	ounces (troy)	14.5833
pounds	pounds (troy)	1.21528
quarts (dry)	cubic inches	67.20
quarts (liq.)	cubic inches	57.75
quarts (liq.)	gallons	0.25
quarts (liq.)	liters	0.9463
radians	degrees	57.30
radians	minutes	3438.0
revolutions	degrees	360.0
revolutions/min	degrees/sec	6.0
rods	meters	5.029
rods	feet	16.5
rods (surveyor's measure)	yards	5.5
seconds	minutes	1.667 EE–2
slugs	pounds	32.17
tons (long)	kilograms	1016.0
tons (short)	kilograms	907.1848
tons (long)	pounds	2240.0
tons (short)	pounds	2000.0
tons (long)	tons (short)	1.120
tons (short)	tons (long)	0.89287
volt (absolute)	statvolts	3.336 EE–3
watts	BTU's/hour	3.4129
watts	horsepower	1.341 EE–3
yards	meters	0.9144
yards	miles (nautical)	4.934 EE–4
yards	miles (statute)	5.682 EE–4

FOR INSTANT RECALL

FUNDAMENTAL CONSTANTS AND UNITS

Avogadro's number: 6.023 EE23 molecules/gmole
Universal gas constant: 1545 ft-lbf/°R-pmole
Speed of light: 3 EE8 m/s
g_c: 32.2 lbm-ft/lbf-sec^2
J, Joules' constant: 778 ft-lbf/BTU

CONVERSIONS

multiply	by	to obtain
atmospheres	29.92	inches of mercury
BTU's	778	ft-lbf
cubic feet	7.48	gallons
gallons	0.1337	cubic feet
gpm	0.00223	cubic feet/sec
horsepower	33,000	ft-lbf/min
	550	ft-lbf/sec
	746	watts
kilowatts	737.6	ft-lbf/sec
microns	EE–6	meters
mph	1.467	ft/sec
slugs	32.2	lbm

TEMPERATURE CONVERSIONS

$$°F = 32 + \left(\tfrac{9}{5}\right)°C$$
$$°C = \tfrac{5}{9}(°F - 32)$$
$$°R = °F + 460$$
$$°K = °C + 273$$
$$\Delta°R = \left(\tfrac{9}{5}\right)\Delta°K$$
$$\Delta°K = \left(\tfrac{5}{9}\right)\Delta°R$$

SI CONVERSIONS

multiply	by	to obtain
kPa	0.145	psi
kg	0.068522	slug
N	0.2248	lbf

DERIVED DATA, PHYSICAL PROPERTIES, AND UNITS

atmospheric pressure: 14.7 psia, 29.92 inches of mercury
circle: 360 degrees, 2π radians
density of air at 1 atmosphere and 70 °F: 0.075 lbm/ft^3
density of mercury: 0.491 lbm/in^3
density of water: 62.4 lbm/ft^3, 0.0361 lbm/in^3, 1 g/cm^3

specific gravity:		
	mercury	13.6
	water	1.0

frequency of house current: 60 Hz
standard gravity: 32.2 ft/sec^2, 386 in/sec^2
modulus of elasticity for steel: 3 EE7 psi
modulus of shear for steel: 1.2 EE7 psi

molecular weight:		
	air	29.0
	carbon	12.0
	carbon dioxide	44.0
	helium	4.0
	hydrogen	2.0
	nitrogen	28.0
	oxygen	32.0

ratio of specific heats for air: 1.4
specific gas constant for air: 53.3 ft-lbf/°R-lbm

specific heat:		
	ice	0.5 (approx.)
	water	1.0
	steam	0.5 (approx.)

triple point of water: 32.02 °F, 0.0888 psia

FORMULAS

area of a circle	πr^2
circumference of a circle	$2\pi r$
area of a triangle	$\frac{1}{2}bh$
volume of a sphere	$\frac{4}{3}\pi r^3$

area moment of inertia for a rectangle

$$I_{centroid} = \tfrac{1}{12}bh^3 \quad I_{side} = \tfrac{1}{3}bh^3$$

area moment of inertia for a circle

$$I_{centroid} = \tfrac{1}{4}\pi r^4$$

polar moment of inertia for a circle

$$J = \tfrac{1}{2}\pi r^4$$

mass moment of inertia of cylinder

$$J = \tfrac{1}{2}mr^2$$

pressure and head

$$p = \rho h$$

rotational speed

$$\omega = 2\pi f = 2\pi\left(\frac{rpm}{60}\right)$$

decibels

$$dB = 10\log_{10}\left(\frac{P_2}{P_1}\right)$$

constant	SI	English
charge on electron	−1.602 EE−19 C	
charge on proton	+1.602 EE−19 C	
atomic mass unit	1.66 EE−27 kg	
electron rest mass	9.11 EE−31 kg	
proton rest mass	1.673 EE−27 kg	
neutron rest mass	1.675 EE−27 kg	
earth mass	6.00 EE24 kg	4.11 EE23 slug; 1.32 EE25 lbm
mean earth radius	6.37 EE3 km	2.09 EE7 ft
mean earth density	5.52 EE3 kg/m^3	3.45 lbm/ft^3
earth escape velocity	1.12 EE4 m/s	3.67 EE4 ft/sec
distance from sun	1.49 EE11 m	4.89 EE11 ft
Boltzmann constant	1.381 EE−23 J/°K	5.65 EE−24 $\frac{\text{ft-lbf}}{°\text{R}}$
permeability of a vacuum	1.257 EE−6 H/m	
permittivity of a vacuum	8.854 EE−12 F/m	
Planck's constant	6.626 EE−34 J·s	
Avogadro's number	6.023 EE23 molecules/gmole	2.73 EE26 $\frac{\text{molecules}}{\text{pmole}}$
Faraday's constant	9.648 EE4 C/gmole	
Stephan-Boltzmann constant	5.670 EE−8 $W/m^2{\cdot}K^4$	1.71 EE−9 $\frac{\text{BTU}}{\text{ft}^2\text{-hr-}°\text{R}^4}$
gravitational constant (G)	6.672 EE−11 $m^3/s^2{\cdot}kg$	3.44 EE−8 $\frac{\text{ft}^4}{\text{lbf-sec}^4}$
universal gas constant	8.314 J/°K·gmole	1545 $\frac{\text{ft-lbf}}{°\text{R-pmole}}$
speed of light	3.00 EE8 m/s	9.84 EE8 ft/sec
speed of sound, air, STP	3.31 EE2 m/s	1.09 EE3 ft/sec
speed of sound, air, 70°F, one atmosphere	3.44 EE2 m/s	1.13 EE3 ft/sec
standard atmosphere	1.013 EE5 N/m^2	14.7 psia
standard atmosphere	0 °C	32 °F
molar ideal gas volume (STP)	22.4138 EE−3 m^3/gmole	359 ft^3/pmole
standard water density	1 EE3 kg/m^3	62.4 lbm/ft^3
air density, STP	1.29 kg/m^3	8.05 EE−2 lbm/ft^3
air density, 70°F, 1 atm	1.20 kg/m^3	7.49 EE−2 lbm/ft^3
mercury density	1.360 EE4 kg/m^3	8.49 EE2 lbm/ft^3
gravity on moon	1.67 m/s^2	5.47 ft/sec^2
gravity on earth	9.81 m/s^2	32.17 ft/sec^2

VOLUME AND SURFACE AREAS

A = cross sectional area or end area
S = lateral surface area
V = enclosed volume

RIGHT CIRCULAR CYLINDER

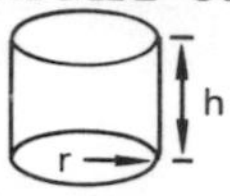

$A = \pi r^2$
$S = 2\pi rh$
$V = Ah = \pi r^2 h$

SPHERE

$S = 4\pi r^2$
$V = \frac{4}{3}\pi r^3$

RIGHT CIRCULAR CONE

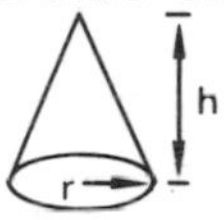

$A_{base} = \pi r^2$
$S = \pi r\sqrt{r^2 + h^2}$
$V = \frac{1}{3}\pi r^2 h$

QUADRATIC EQUATION

The roots of the equation $ax^2 + bx + c = 0$ are

$$\frac{-b \pm \sqrt{b^2 - 4ac}}{2a}$$

DETERMINANT

A determinant of order n is a square array of n^2 values enclosed within vertical bars. Each determinant has a certain value. For a second-order determinant:

$$\begin{vmatrix} a_1 & a_2 \\ b_1 & b_2 \end{vmatrix} = a_1 b_2 - a_2 b_1$$

For a third-order determinant:

$$\begin{vmatrix} a_1 & a_2 & a_3 \\ b_1 & b_2 & b_3 \\ c_1 & c_2 & c_3 \end{vmatrix} = a_1 b_2 c_3 + a_2 b_3 c_1 + a_3 b_1 c_2$$
$$- c_1 b_2 a_3 - c_2 b_3 a_1 - c_3 b_1 a_2$$

ARITHMETIC SERIES

An arithmetic series is given by

$$a + (a + d) + (a + 2d) + \cdots$$

If the series has a finite number of terms, it converges. If the series has an infinite number of terms, it diverges. The sum of n terms is found by consecutive addition. The nth term is

$$b_n = a + (n - 1)d$$
$$B_n = \tfrac{1}{2}n[2a + (n - 1)d]$$

GEOMETRIC SERIES

A geometric series is given by

$$a + aR + aR^2 + \cdots + aR^{n-1}$$

If n is finite, the series converges. The nth term and sum are

$$b_n = aR^{n-1} \qquad B_n = \frac{a(1 - R^n)}{1 - R}$$

If n is infinite, the series converges only for $-1 < R < 1$. In this case, the sum of infinite terms is

$$B_n = \frac{a}{1 - R}$$

The series diverges if $|R| \geq 1$.

TRIGONOMETRY

Trigonometric functions are defined in terms of a right triangle.

$$\sin\theta = \frac{y}{h} \qquad \cos\theta = \frac{x}{h} \qquad \tan\theta = \frac{y}{x}$$

$$\csc\theta = \frac{h}{y} \qquad \sec\theta = \frac{h}{x} \qquad \cot\theta = \frac{x}{y}$$

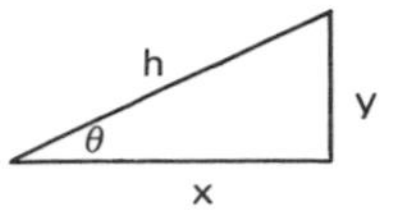

Basic Identities: $\sin^2\theta + \cos^2\theta = 1$

$$\sin 2\theta = 2\sin\theta\cos\theta$$
$$1 + \tan^2\theta = \sec^2\theta$$
$$\cos 2\theta = \cos^2\theta - \sin^2\theta = 2\cos^2\theta - 1$$

For general triangles: The *law of sines* is

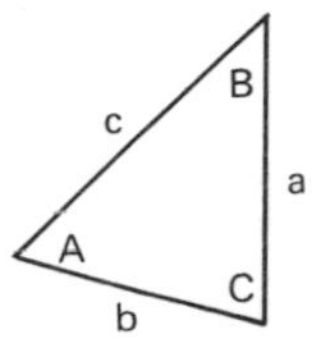

$$\frac{\sin A}{a} = \frac{\sin B}{b} = \frac{\sin C}{c}$$

The *law of cosines* is

$$a^2 = b^2 + c^2 - 2bc\cos A$$

The area of a general triangle is $\frac{1}{2}ab(\sin C)$.

Logarithm Identities:

$x^a = \text{antilog}[a\log(x)]$ — $\ln(x) = (\log_{10}x)/(\log_{10}e)$
$\log(x^a) = a\log(x)$ — $\approx 2.3(\log_{10}x)$
$\log(xy) = \log(x) + \log(y)$ — $\log_b(b) = 1$
$\log(x/y) = \log(x) - \log(y)$ — $\log(1) = 0$
$\log_b(b^n) = n$

COMPLEX NUMBERS (ϕ in radians)

$a + ib \equiv r\angle\phi \equiv r[\cos\phi + i\sin\phi] \equiv re^{i\phi}$
$a - ib \equiv r\angle\phi \equiv r[\cos\phi - i\sin\phi] \equiv re^{-i\phi}$
$r = \sqrt{a^2 + b^2}$
$\phi = \arctan\left(\frac{b}{a}\right)$

EQUATIONS OF A STRAIGHT LINE

The general form is $Ax + By + C = 0$.

The slope-intercept form is $y = mx + b$ where m is the slope and b is the y-intercept.

Intercept form: $\frac{x}{a} + \frac{y}{b} = 1$

Slopes of perpendicular lines: $m_2 = \frac{-1}{m_1}$

Given one point on the line (x^*, y^*), the point-slope form is $y - y^* = m(x - x^*)$.

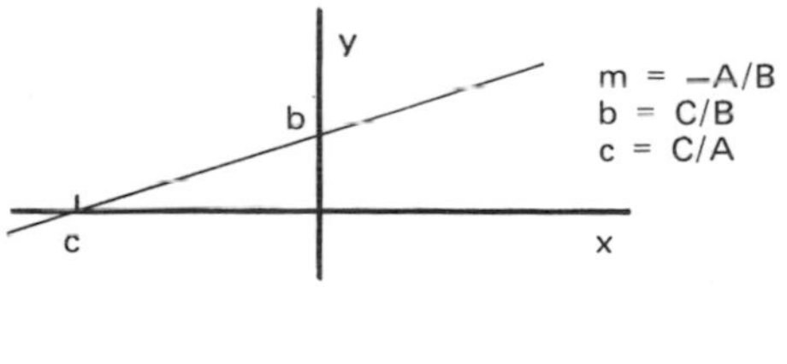

CONIC SECTIONS

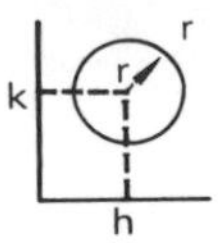

If the cutting plane is perpendicular to the axis of the cone, the intersection is a circle. The equation of a circle of radius r which is centered at (h, k) is

$$(x-h)^2+(y-k)^2=r^2$$

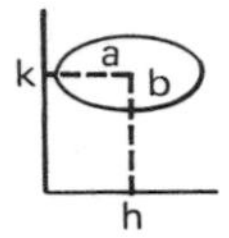

If the cutting plane is inclined to the axis of the cone, the intersection will be an ellipse. The equation of an ellipse centered at (h, k) with a and b defined as shown is

$$\left(\frac{x-h}{a}\right)^2+\left(\frac{y-k}{b}\right)^2=1$$

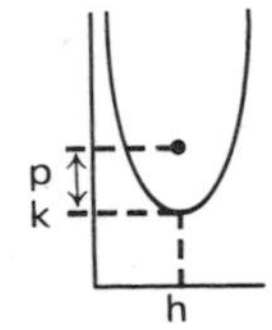

If the cutting plane is parallel to the cone surface, the intersection is a parabola. The equation of a parabola with vertex at (h, k) and symmetric with respect to the y-axis is

$$(x-h)^2=4p(y-k)$$

EXTREMA BY DIFFERENTIATION

Given a continuous function, $f(x)$, the extreme points can be found by taking the first derivative and setting it equal to zero. Let x^* be the value of x which satisfies this equality. $f(x^*)$ is a minimum if $f''(x^*)$ is greater than zero. If $f''(x^*)$ is less than zero, $f(x^*)$ is a maximum. $f''(x^*)$ is equal to zero at an *inflection point.*

VECTORS

A vector is a directed line segment. It is defined completely only when both the magnitude and direction are known. Vectors are defined in terms of the unit vectors **i, j, k**.

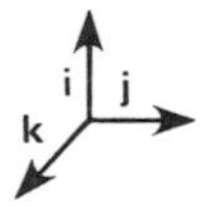

The unit vectors are vectors of length one directed along the x, y, and z axes, respectively. A vector, **A**, can be written in terms of the unit vectors and its endpoints (a_x, a_y, a_z)

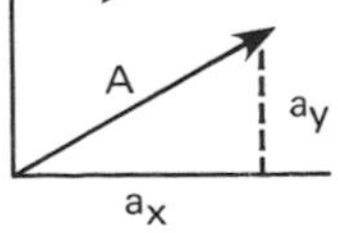

$$\mathbf{A}=a_x\mathbf{i}+a_y\mathbf{j}+a_z\mathbf{k}$$

Addition of vectors is performed by adding the components:

$$\mathbf{A}+\mathbf{B}=(a_x+b_x)\mathbf{i}+(a_y+b_y)\mathbf{j}+(a_z+b_z)\mathbf{k}$$

The *dot product* is a scalar and represents the projection of **B** onto **A**. It is given by

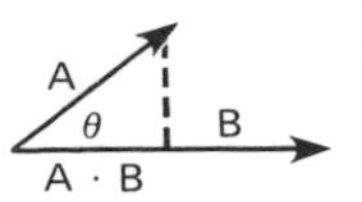

$$\mathbf{A}\cdot\mathbf{B}=a_xb_x+a_yb_y=|\mathbf{A}||\mathbf{B}|\cos\theta$$

The *cross product* is a vector of magnitude $|\mathbf{B}||\mathbf{A}|\sin\theta$ which is perpendicular to the plane containing **A** and **B**. The product is

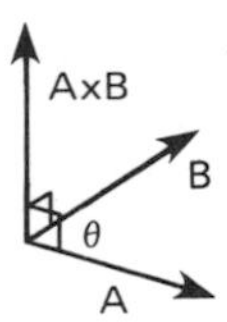

$$\mathbf{A}\times\mathbf{B}=\begin{vmatrix}\mathbf{i}&\mathbf{j}&\mathbf{k}\\a_x&a_y&a_z\\b_x&b_y&b_z\end{vmatrix}$$

The sense of $\mathbf{A}\times\mathbf{B}$ is determined by the right-hand rule.

MENSURATION

CIRCLE of radius r:

$$\text{circumference}=2\pi r=p$$

$$\text{area}=\pi r^2=\frac{p^2}{4\pi}$$

ELLIPSE with axes a and b:

$$\text{area}=\pi ab$$

CIRCULAR SECTOR with radius r, includ-ed angle θ, and arc length s:

$$\text{area}=\tfrac{1}{2}\theta r^2=\tfrac{1}{2}sr$$

$$\text{arc length}=s=\theta r$$

The angle θ is in radians.

PERMUTATIONS AND COMBINATIONS

$$P(n,r)=\frac{n!}{(n-r)!}$$

$$C(n,r)=\frac{n!}{r!(n-r)!}$$

BINOMIAL PROBABILITY DISTRIBUTION

$f(x)$ is the probability that x will occur in n trials.
p is the probability of one success in one trial.
$q=(1-p)$ is the probability of one failure in one trial.

$$f(x)=\binom{n}{x}p^xq^{n-x}=\frac{n!}{(n-x)!x!}\,p^xq^{n-x}$$

POISSON PROBABILITY DISTRIBUTION

$f(x)$ is the probability of x occurrences. x is the actual number of occurences in some period. λ is the mean number of occurences per period.

$$f(x)=\frac{e^{-\lambda}\lambda^x}{x!}$$

STATISTICS

$$\text{arithmetic mean}=\bar{x}=\left(\tfrac{1}{n}\right)(x_1+x_2+\cdots x_n)=\frac{\sum x_i}{n}$$

$$\text{geometric mean}=\sqrt[n]{x_1x_2x_3\cdots x_n}$$

$$\text{harmonic mean}=\frac{n}{\frac{1}{x_1}+\frac{1}{x_2}+\cdots+\frac{1}{x_n}}$$

$$\text{root-mean-squared value}=\sqrt{\frac{\sum x_i^2}{n}}$$

$$\text{standard deviation}=\sigma=\sqrt{\frac{\sum(x_i-\bar{x})^2}{n}}=\sqrt{\frac{\sum x_i^2}{n}-(\bar{x})^2}$$

$$\text{sample standard deviation}=s=\sqrt{\frac{\sum(x_i-\bar{x})^2}{n-1}}=\sqrt{\frac{\sum x_i^2-\frac{\left(\sum x_i\right)^2}{n}}{n-1}}$$

NOMENCLATURE AND DEFINITIONS

A	Annual amount or annuity: a payment or receipt of a fixed sum of money at yearly intervals.
	Amortization: the setting aside of money at intervals, as in a sinking fund, for gradual payment of a debt; or, the writing off of capital investments by prorating initial cost over a fixed period.
B	Book value: the theoretical market value of an asset at any time within its useful life. Book value is calculated as the initial cost less any accumulated depreciation.
C	Cost: the asset purchase price.
d	Declining balance depreciation rate: For double declining balance depreciation, $d = 2/n$.
D_j	Depreciation in year j: the allowance in year j of the life of the asset for the decrease in value due to wear, deterioration, and obsolescence.
EUAC	Equivalent Uniform Annual Cost: the present worth of an alternative multiplied by the (A/P) factor.
F	Future worth, value, or amount: the value of an asset at some future point in time. May be calculated by multiplying the present worth of an alternative by the (F/P) factor.
G	Uniform gradient amount: a quantity by which the cash flows change each year.
i	Annual effective interest rate.
k	Number of compounding periods per year.
MARR	Minimum Attractive Rate of Return (usually the same as the annual effective interest rate, i).
n	Number of compounding periods; or the expected life of an asset.
P	Present worth, value, or amount. The value of an amount at time = 0.
	Nominal annual interest rate (same as rate per annum).
ROR	Rate of Return: the interest rate which makes the present worth equal to zero.
ROI	Return On Investment.
S_n	Expected salvage value in year n.
t	Time or income tax rate.
ϕ	Effective interest rate per period (equal to $\frac{r}{k}$).

YEAR-END CONVENTION

All cash disbursements and receipts (cash flows) are assumed to occur at the end of the year in which they actually occur. The exception is an initial cost (purchase price) which is assumed to occur at time = 0.

SUNK COSTS

Sunk costs are expenses incurred before time = 0. They have no bearing on the evaluation of alternatives.

CASH FLOW DIAGRAMS

Cash flow diagrams are drawn to help visualize problems involving transfers of money at various points in time. The following conventions are used in their construction:

(a) The horizontal axis represents time. The axis is divided into equal increments, one per period.

(b) The year-end convention is assumed.

(c) Disbursements are downward arrows; receipts are upward arrows. In both cases, the length of the arrow is proportional to the magnitude of the transfer.

(d) Two or more transfers in the same time period are represented by end-to-end arrows.

DISCOUNTING FACTORS

Discounting factors are numbers used to calculate the equivalent amount (at some point in time) of an alternative. The factors are given the functional notation $(X, Y, i\%, n)$ where X is the desired value, Y is the known value, i is the interest rate, and n is the number of periods. (Usually, n is in years). The functional symbol is read "... X given Y at $i\%$ for n years ...". The discounting factors convert Y to X under the conditions of constant interest rate compounded for n years. The common discounting factors (for discrete compounding) are given below:

single payment compound amount	$(F/P, i\%, n)$	$(1+i)^n$
present worth	$(P/F, i\%, n)$	$(1+i)^{-n}$
uniform series sinking fund	$(A/F, i\%, n)$	$\dfrac{i}{(1+i)^n - 1}$
capital recovery	$(A/P, i\%, n)$	$\dfrac{i(1+i)^n}{(1+i)^n - 1}$
compound amount	$(F/A, i\%, n)$	$\dfrac{(1+i)^n - 1}{i}$
equal series present worth	$(P/A, i\%, n)$	$\dfrac{(1+i)^n - 1}{i(1+i)^n}$
uniform gradient	$(P/G, i\%, n)$	$\left(\dfrac{1}{i} - \dfrac{n}{(1+i)^n - 1}\right) \times (P/A, i\%, n)$

COMPARING ALTERNATIVES

The PRESENT WORTH method may be used if all of the alternatives have equal lives. The present worth of each alternative is calculated. The alternative with the smallest negative or largest positive present worth is chosen.

The EQUIVALENT UNIFORM ANNUAL COST (EUAC) method must be used if alternatives have unequal lives. Use of this method is restricted to alternatives which are infinitely renewed—specifically, each alternative is replaced by an identical replacement at the end of its useful life, up to the duration of the longest-lived alternative. The EUAC of an alternative is usually found from the following formula:

$$\text{EUAC} = \begin{pmatrix}\text{present worth} \\ \text{of alternative}\end{pmatrix} (A/P, i\%, n)$$

The CAPITALIZED COST of a project is the present worth of a project which has an infinite life. The capitalized cost represents the amount of money needed at time = 0 to support the project on interest only, without reducing the principal. For annual operating and maintenance costs:

$$\text{capitalized cost} = \text{inital cost} + \frac{\text{annual maintenance cost}}{i}$$

For a project with maintenance costs every n years,

$$\text{capitalized cost} = \text{initial cost} + (\text{maintenance cost})\frac{(A/P, i\%, n)}{i}$$

The BENEFIT/COST RATIO is

$$\text{B/C} = \frac{\text{pw of benefits} - \text{pw of disbenefits}}{\text{pw of costs}}$$

RATE OF RETURN

The rate of return (ROR) is the interest rate that makes the present worth equal to zero. To determine the ROR, assume a reasonable value for the interest rate earned and find the present worth. If the present worth is zero, the assumed interest rate is the ROR. If the present worth is not zero, assume another value of i and find the present worth. Use the two values of i and their corresponding present worths to interpolate or extrapolate a value of i which makes the present worth zero.

NON-ANNUAL COMPOUNDING

For problems in which compounding is done at intervals other than yearly, an effective annual interest rate can be computed from the following formula:

$$i = \left(1 + \frac{r}{k}\right)^k - 1 = (1+\phi)^k - 1$$

$$\phi = \frac{r}{k} = \text{effective rate per period}$$

DEPRECIATION

STRAIGHT LINE $\quad D_j = \dfrac{C - S_n}{n}$

DOUBLE DECLINING BALANCE ($i = 1$ to $j-1$) $\quad D_j = \dfrac{2(C - \sum D_i)}{n}$

$$D_j = \frac{2C}{n}\left(1 - \frac{2}{n}\right)^{j-1}$$

$$BV_j = C\left(1 - \frac{2}{n}\right)^j$$

SUM OF THE YEAR'S DIGITS $\quad D_j = \dfrac{(C - S_n)(n - j + 1)}{T}$

$$T = \tfrac{1}{2}n(n+1)$$

SINKING FUND $\quad D_j = (C - S_n)(A/F, i\%, n) \times (F/P, i\%, j-1)$

ACRS $\quad D_j = (\text{factor})C$

BOOK VALUE

$$\text{BV} = \text{initial cost} - \sum D_j$$

INCOME TAXES

If income taxes are paid, operating expenses and depreciation are deductible. If t is the tax rate, revenues and all expenses (except depreciation) should be multiplied by $(1 - t)$ in the year in which they occur. Although depreciation is a deductible expense, it is not an actual out-of-pocket expense. Depreciation should be multiplied by t and added to the cash flow in the appropriate year.

CONSUMER LOANS

BAL_j	balance after the jth payment
LV	total value loaned (cost − down payment)
j	payment or period number
N	total number of payments to pay off the loan
PI_j	jth interest payment
PP_j	jth principal payment
PT_j	jth total payment
ϕ	effective rate per period ($\frac{r}{k}$)

SIMPLE INTEREST

Interest due does not compound with a *simple interest* loan. The interest due is proportional to the length of time the principal is outstanding.

DIRECT REDUCTION LOANS

This is the typical "interest paid on unpaid balance" loan. The amount of the periodic payment is constant, but the amounts paid towards the principal and interest both vary.

$$N = \frac{-\ln\left(\dfrac{-\phi(LV)}{PT} + 1\right)}{\ln(1+\phi)}$$

$$PT = LV(A/P, i\%, n)$$

$$LV = \frac{PT}{(A/P, i\%, n)}$$

$$BAL_{j-1} = PT\left(\frac{1 - (1+\phi)^{j-1-N}}{\phi}\right)$$

$$PI_j = \phi(BAL_{j-1})$$

$$PP_j = PT - PI_j$$

$$BAL_j = BAL_{j-1} - PP_j$$

HANDLING INFLATION

It is important to perfom economic studies in terms of *constant value dollars.* One method of converting all cash flows to constant value dollars is to divide the flows by some annual *economic indicator* or price index.

An alternative is to replace i with a value corrected for the inflation rate, e. This corrected value, i', is

$$i' = i + e + ie$$

DENSITY AND SPECIFIC GRAVITY

water: 62.4 lbm/ft^3, 0.0361 lbm/in^3, 8.345 lbm/gal; SG = 1.0
mercury: 848.4 lbm/ft^3, 0.491 lbm/in^3; SG = 13.6
air: 0.075 lbm/ft^3 at STP

VISCOSITY

μ: absolute (dynamic) viscosity (lbf-sec/ft^2)

$$= \frac{\text{shear stress}}{\text{rate of shear strain}}$$

ν: kinematic viscosity ($\text{ft}^2\text{/sec}$) $= \mu g_c/\rho$

HYDROSTATIC PRESSURE

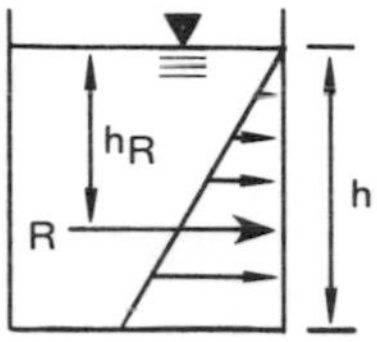

$$p = \rho h$$
$$\bar{p} = \tfrac{1}{2}\rho h$$
$$R = \bar{p}A = \tfrac{1}{2}\rho h A$$
$$h_R = \tfrac{2}{3}h$$

PRESSURE ON SUBMERGED PLANE SURFACES

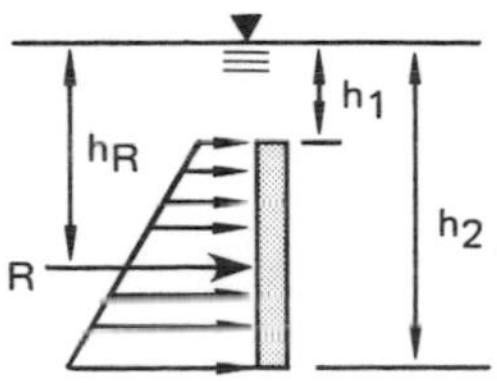

h_c = distance from surface to centroid of plane, as measured parallel to plane's surface

$$\bar{p} = \rho h_c = \tfrac{1}{2}\rho(h_1 + h_2)$$
$$R = \bar{p}A$$
$$h_R = h_c + \frac{I_c}{Ah_c}$$ as measured parallel to plane's surface

MANOMETERS

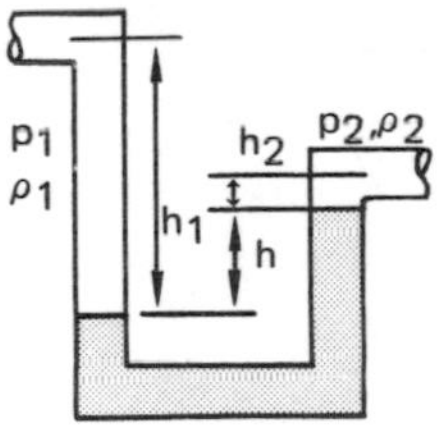

$$\Delta p = p_1 - p_2 = \rho_m h - \rho_1 h_1 + \rho_2 h_2$$

BUOYANCY (ARCHIMEDES' PRINCIPLE)

The buoyant force is equal to the weight of the displaced fluid.

$$F_{\text{buoyant}} = \rho V_{\text{displaced}}$$

$$V_{\text{displaced}} = \frac{\text{dry weight} - \text{submerged weight}}{\rho}$$

SPEED OF SOUND IN A FLUID

$$c = \sqrt{kg_cRT} = \sqrt{kg_cp/\rho} \quad \text{(gases)}$$

$$c = \sqrt{\frac{Eg_c}{\rho}} \quad \text{(liquids)}$$

$$\text{E (bulk modulus)} = \frac{1}{\text{compressibility}}$$

$\approx$ 4.3 EE7 lbf/ft^2 for water

$$N_{\text{M}} \text{ (Mach number)} = \frac{v}{c}$$

CONTINUITY EQUATION

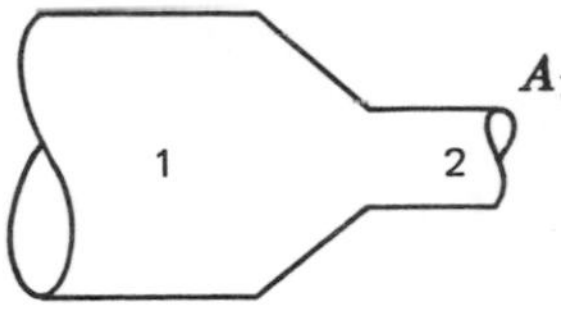

$$\dot{m}_1 = \dot{m}_2$$
$$A_1v_1\rho_1 = A_2v_2\rho_2 \text{ (gases and liquids)}$$
$$A_1v_1 = A_2v_2 \text{ (liquids only)}$$
$$Av\rho = \dot{m}$$
$$Av = Q$$

BERNOULLI EQUATION

$$\frac{p}{\rho} + \frac{v^2}{2g_c} + z = \text{constant}$$

$\frac{p}{\rho}$ = pressure (static) head

$\frac{v^2}{2g_c}$ = velocity (dynamic) head

z = potential (gravitational) head

CONSERVATION OF ENERGY

$$\frac{p_1}{\rho} + \frac{v_1^2}{2g_c} + z_1 + W_{\text{pump}} = \frac{p_2}{\rho} + \frac{v_2^2}{2g_c} + z_2 + h_f + W_{\text{turbine}}$$

W_{pump} and W_{turbine} are on a per-pound basis with units of ft-lbf/lbm.

REYNOLDS NUMBER (dimensionless)

$$N_{\text{Re}} = \frac{vD\rho}{\mu g_c} = \frac{vD}{\nu} = \frac{\text{inertial force}}{\text{viscous force}}$$

$$D = 4r_H$$

laminar 2000 > N_{Re} > 3000 turbulent

FRICTION LOSS (DARCY)

$$h_f = \frac{fLv^2}{2Dg_c} = \frac{Kv^2}{2g_c}$$

K is the friction loss coefficient for fittings

$$f = \frac{64}{N_{\text{Re}}} \text{ for } N_{\text{Re}} < 2000$$

HYDRAULIC HORSEPOWER (W_{pump})

	Q in gpm	$\dot{m}$ in lbm/sec	$\dot{V}$ in cfs
h_A is added head in feet	$\frac{h_AQ(S.G.)}{3956}$	$\frac{h_A\dot{m}}{550}$	$\frac{h_A\dot{V}(S.G.)}{8.814}$
p is added head in psf	$\frac{pQ}{2.468\ \text{EE5}}$	$\frac{p\dot{m}}{(34,320)(S.G.)}$	$\frac{p\dot{V}}{550}$

TORRICELLI EQUATION

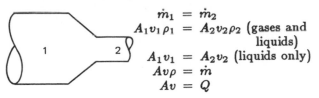

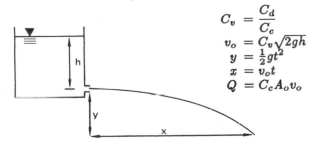

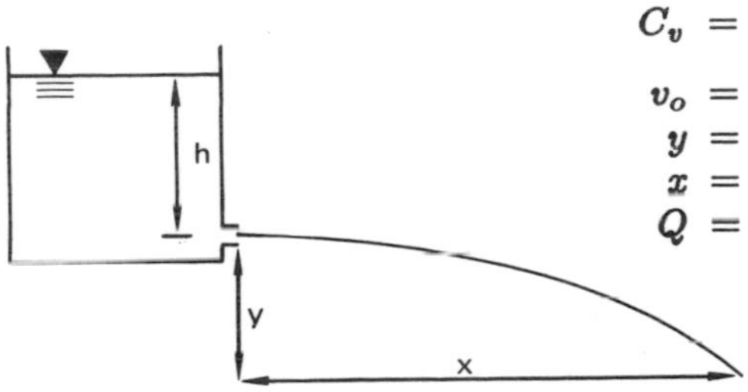

$$C_v = \frac{C_d}{C_c}$$
$$v_o = C_v\sqrt{2gh}$$
$$y = \tfrac{1}{2}gt^2$$
$$x = v_ot$$
$$Q = C_cA_ov_o$$

STATIC-PITOT TUBE

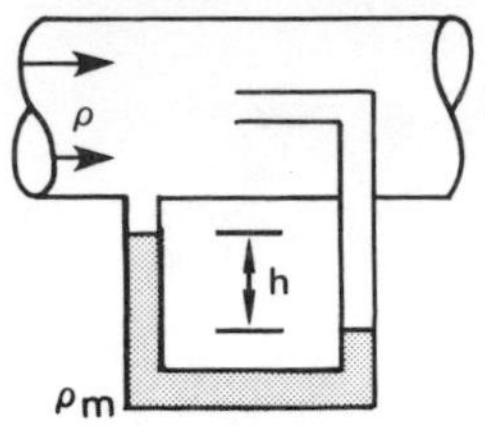

$$v = \sqrt{2g\frac{(\rho_m - \rho)h}{\rho}}$$

$$\rho \approx 0 \text{ for air}$$

VENTURI METER

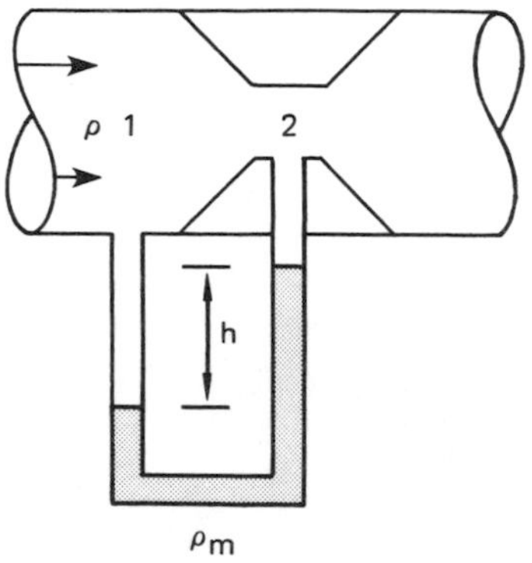

$$\beta = \frac{D_2}{D_1}$$

$$F_{va} = \frac{1}{\sqrt{1-\beta^4}}$$

$$v_2 = F_{va}\sqrt{\frac{2gh(\rho_m - \rho)}{\rho}}$$

$$= F_{va}\sqrt{\frac{2g(p_1 - p_2)}{\rho}}$$

$$C_d = C_c C_v$$

$$C_c \approx 1 \text{ for venturi meters}$$

$$C_f = C_d F_{va}$$

$$Q = C_f A_2 \sqrt{2gh\frac{(\rho_m - \rho)}{\rho}}$$

$$\rho = 0 \text{ for air}$$

ORIFICE PLATE

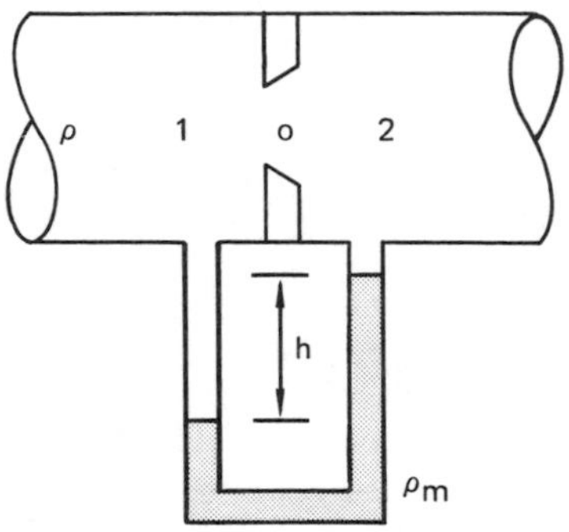

$$F_{va} = \frac{1}{\sqrt{1 - C_c^2\left(\frac{A_o}{A_1}\right)^2}}$$

$$v_2 = F_{va} C_v \sqrt{2gh\frac{(\rho_m - \rho)}{\rho}}$$

$$C_f = F_{va} C_d$$

$$C_d = C_c C_v$$

$$Q = C_c A_o v_2$$

$$= A_o C_f \sqrt{2gh\frac{(\rho_m - \rho)}{\rho}}$$

Q is in cfs

IMPULSE-MOMENTUM

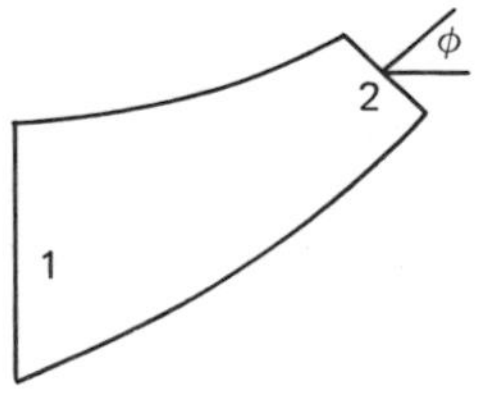

Q is in cfs

$$\dot{m} = \frac{Q\rho}{g_c}$$

$$R_x = A_1 p_1 - A_2 p_2 \cos\phi - \dot{m}(v_2 \cos\phi - v_1)$$

$$R_y = A_2 p_2 \sin\phi + \dot{m} v_2 \sin\phi$$

p's are gage pressures

LIFT AND DRAG

$$L = \frac{C_L v^2 \rho A}{2g_c}$$

$$D = \frac{C_D v^2 \rho A}{2g_c}$$

A is the projected area (a circle for a sphere).

SIMILARITY

For fans, pumps, turbines, drainage through holes in tanks, closed-pipe flow with no free surfaces (in the turbulent region with the same relative roughness), and for completely submerged objects such as torpedoes, airfoils, and submarines, performance of models and prototypes can be correlated by equating Reynolds numbers.

$$(N_{\text{Re}})_{\text{model}} = (N_{\text{Re}})_{\text{prototype}}$$

OPEN CHANNEL FLOW

$$r_H = \text{hydraulic radius} = \frac{\text{area in flow}}{\text{wetted perimeter}} = \frac{D_e}{4}$$

Chezy Equation

$$v = C\sqrt{r_H S}$$

$$S = \text{slope} = h_f/L$$

Manning Equation (omit 1.49 if r_H is in meters)

$$C = \left(\tfrac{1}{n}\right)(1.49)(r_H)^{1/6}$$

$$n = \text{Manning constant}$$

Diameter of a pipe flowing full

$$D = 1.33\left(\frac{Qn}{\sqrt{S}}\right)^{3/8}$$

Diameter of a pipe flowing half-full

$$D = 1.73\left(\frac{Qn}{\sqrt{S}}\right)^{3/8}$$

Hazen-Williams Equation

$$v = (1.318)C^*(r_H)^{0.63}(S)^{0.54}$$

$$C^* = \text{Hazen-Williams coefficient}$$

Froude Number

$$N_{\text{Fr}} = \frac{v}{\sqrt{g \times \text{depth}}}$$

AFFINITY/SIMILARITY LAWS

$$\frac{Q_2}{Q_1} = \frac{n_2}{n_1}$$

$$\frac{H_2}{H_1} = \left(\frac{n_2}{n_1}\right)^2 = \left(\frac{Q_2}{Q_1}\right)^2$$

$$\frac{\text{bhp}_2}{\text{bhp}_1} = \left(\frac{n_2}{n_1}\right)^3 = \left(\frac{Q_2}{Q_1}\right)^3$$

$$\frac{Q_2}{Q_1} = \frac{d_2}{d_1}$$

$$\frac{H_2}{H_1} = \left(\frac{d_2}{d_1}\right)^2$$

$$\frac{\text{bhp}_2}{\text{bhp}_1} = \left(\frac{d_2}{d_1}\right)^3$$

THE ZEROETH LAW OF THERMODYNAMICS

If two bodies are in thermal equilibrium with a third body, then the two bodies are in thermal equilibrium with each other.

THE FIRST LAW OF THERMODYNAMICS

One form of energy may be converted into another form, but energy cannot be created or destroyed.

THE SECOND LAW OF THERMODYNAMICS

Heat will not flow by itself from a cold body to a hot body. Entropy always increases.

THE THIRD LAW OF THERMODYNAMICS

The entropy of a perfect substance (crystal) at absolute zero is zero.

TYPES OF THERMODYNAMIC SYSTEMS

A *system* is a region with artificially-chosen boundaries. A *closed system* is one in which matter does not cross the system boundaries. Energy may cross the system boundaries, however. Both energy and matter may cross the boundaries for an *open system*.

MATHEMATICAL FORMULATION OF THE FIRST LAW

For closed systems:

$$dQ = dU + dW \quad \text{or} \quad Q = \Delta U + W$$

For steady-flow open systems:

$$Q = W + h_2 - h_1 + \frac{z_2 - z_1}{J} + \frac{v_2^2 - v_1^2}{2g_c J}$$

(Q, W, and h are on a per-pound basis)

TEMPERATURE SCALES

$$°\text{R} = °\text{F} + 460$$

$$°\text{K} = °\text{C} + 273$$

$$°\text{F} = 32 + \left(\tfrac{9}{5}\right)°\text{C}$$

$$°\text{C} = \left(\tfrac{5}{9}\right)(°\text{F} - 32)$$

$$\Delta°\text{R} = \left(\tfrac{9}{5}\right)\Delta°\text{K}$$

$$\Delta°\text{K} = \left(\tfrac{5}{9}\right)\Delta°\text{R}$$

IMPORTANT CONVERSIONS

$$1 \text{ BTU} = 778 \text{ ft-lbf} \qquad \left(J = 778 \, \frac{\text{ft-lbf}}{\text{BTU}}\right)$$

$$1 \text{ BTU} = 252 \text{ calories}$$

$$1 \text{ hp} = 550 \text{ ft-lbf/sec} = 2545 \text{ BTU/hr}$$

$$1 \text{ kw} = 3412.9 \text{ BTU/hr}$$

$$1 \text{ kw} = 1.341 \text{ hp}$$

STP (STANDARD TEMPERATURE AND PRESSURE)

Scientific STP is 32 °F and 14.7 psia.
For fuel gases, STP is 60 °F and 14.7 psia.

COMPOSITION OF DRY ATMOSPHERIC AIR

	% by weight	% by volume
oxygen (O_2)	23.15	20.9
nitrogen (N_2)	76.85	79.1

(rare inert gases included as N_2)

IDEAL GAS EQUATION OF STATE

$$pV = nR^*T \qquad \text{(for n moles)}$$

$$pV = mRT \qquad \text{(for m pounds mass)}$$

$$R^* = 1545 \text{ ft-lbf/pmole-}°\text{R}$$

$$R^* = 0.08206 \text{ atm-liters/gmole-}°\text{K}$$

$$R = 53.3 \text{ ft/}°\text{R for air}$$

COMPRESSED GAS EQUATION OF STATE

$$pV = mZRT$$

$$Z = \text{compressibility factor}$$

IDEAL GAS PROCESS LAW

$$\frac{p_1 V_1}{T_1} = \frac{p_2 V_2}{T_2}$$

SPECIFIC HEAT RELATIONSHIPS FOR IDEAL GASES

$$Q = mc\Delta T$$

$$c_p = 0.24 \text{ BTU/lbm-}°\text{F} \qquad \text{(for air)}$$

$$c_v = 0.1714 \text{ BTU/lbm-}°\text{F} \qquad \text{(for air)}$$

$$k = \frac{c_p}{c_v} = 1.4 \qquad \text{(for air)}$$

$$c_p - c_v = \frac{R}{J}$$

$$c_p = \frac{Rk}{J(k-1)}$$

MIXTURES OF IDEAL GASES

Volumetrically weighted: density and molecular weight of the mixture

Gravimetrically weighted: internal energy, enthalpy, entropy, specific heats, and specific gas constant of the mixture

MISCELLANEOUS GAS LAWS

Avogadro: Equal volumes of different gases under the same conditions of temperature and pressure contain the same number of molecules (6.023 EE23 molecules per gmole).

Dalton: The total pressure of a mixture of gases is equal to the sum of the partial pressures. (The partial pressures are determined at the final mixture temperature and volume.)

Amagat: The total volume of a mixture of gases is equal to the sum of the partial volumes. (The partial volumes are determined at the final mixture temperature and pressure.)

THERMODYNAMIC RELATIONSHIPS FOR ANY PROCESS (FOR IDEAL GASES)

$\Delta u = c_v \Delta T$ ($c_v = 0.1714$ BTU/lbm-°F for air)
$\Delta h = c_p \Delta T$ ($c_p = 0.24$ BTU/lbm-°F for air)

TYPES OF PROCESSES

Isochoric (Isometric): constant volume
Isobaric: constant pressure
Isothermal: constant temperature
Polytropic: any process for which $p(V)^n$ is constant
Adiabatic: a process with no heat transfer ($Q = 0$)
Isentropic: an adiabatic process for which $\Delta s = 0$
Throttling: an adiabatic process for which $\Delta h = 0$

SENSIBLE HEAT

$Q = mc_p(T_2 - T_1)$
$c_p = 1$ BTU/lbm-°F $= 1$ cal/gram-°C (for water)

LATENT HEAT

Latent heat is the energy which changes the phase of a substance.
Latent heat of fusion: 143.4 BTU/lbm for turning ice to liquid
Latent heat of vaporization: 970.3 BTU/lbm for turning liquid to steam
Latent heat of sublimation: 1293.5 BTU/lbm for turning ice to steam directly
The above values are for water phases at 1 atmosphere.

QUALITY OF A LIQUID-VAPOR MIXTURE

$$x = \frac{\text{mass of vapor}}{\text{mass of liquid} + \text{mass of vapor}}$$

PROPERTIES OF A LIQUID-VAPOR MIXTURE

$$h = h_F + xh_{FG}$$
$$u = u_F + xu_{FG}$$
$$v = v_F + xv_{FG}$$
$$s = s_F + xs_{FG}$$

ATMOSPHERIC AIR DEFINITIONS

The *dry-bulb temperature* is the temperature of still air as measured by a regular thermometer.
The *wet-bulb temperature* is the temperature as measured during an adiabatic-saturation process.
The *dewpoint temperature* is the temperature at which moisture in the air begins to condense out when the air is cooled through a constant-pressure process.
The *humidity ratio (specific humidity)* is the ratio of the mass of water vapor in the air to the mass of the dry air alone.
The *relative humidity* is the ratio of the partial pressure of the vapor to the saturation pressure at the air temperature.

CONSTANT PRESSURE PROCESSES (CLOSED SYSTEM, IDEAL GAS)

$$p_2 = p_1$$
$$T_2 = \frac{T_1 v_2}{v_1}$$
$$v_2 = \frac{v_1 T_2}{T_1}$$
$$Q = \Delta h = c_p(T_2 - T_1)$$
$$u_2 - u_1 = c_v(T_2 - T_1) = \frac{p_2 v_2 - p_1 v_1}{k-1}$$
$$= c_v \frac{p_2 v_2 - p_1 v_1}{R}$$
$$W = p_1(v_2 - v_1) = p_2(v_2 - v_1)$$
$$= R(T_2 - T_1)$$
$$s_2 - s_1 = c_p \ln \frac{T_2}{T_1} = c_p \ln \frac{v_2}{v_1}$$
$$h_2 - h_1 = Q = c_p(T_2 - T_1) = kp_1 \frac{v_2 - v_1}{k-1}$$

CONSTANT TEMPERATURE PROCESSES (CLOSED SYSTEM, IDEAL GAS)

$$p_2 = \frac{p_1 v_1}{v_2}$$
$$T_2 = T_1$$
$$v_2 = \frac{v_1 p_1}{p_2}$$
$$Q = p_1 v_1 \ln \frac{v_2}{v_1} = T(s_2 - s_1)$$
$$u_2 - u_1 = 0$$
$$W = p_1 v_1 \ln \frac{v_2}{v_1} = RT \ln \frac{p_1}{p_2}$$
$$s_2 - s_1 = \frac{Q}{T} = R \ln \frac{v_2}{v_1}$$
$$= R \ln \frac{p_1}{p_2}$$
$$h_2 - h_1 = 0$$

CONSTANT VOLUME PROCESSES (CLOSED SYSTEM, IDEAL GAS)

$$p_2 = \frac{p_1 T_2}{T_1}$$
$$T_2 = \frac{T_1 p_2}{p_1}$$
$$v_2 = v_1$$
$$Q = u_2 - u_1 = c_v(T_2 - T_1)$$
$$u_2 - u_1 = Q = c_v(T_2 - T_1)$$
$$= \frac{p_2 v_2 - p_1 v_1}{k-1}$$
$$W = 0$$
$$s_2 - s_1 = c_v \ln\left(\frac{T_2}{T_1}\right) = c_v \ln\left(\frac{p_2}{p_1}\right)$$
$$h_2 - h_1 = c_p(T_2 - T_1) = \frac{kv_1(p_2 - p_1)}{k-1}$$

ISENTROPIC PROCESSES (CLOSED SYSTEM, IDEAL GAS) (REVERSIBLE ADIABATIC)

$$p_2 = p_1\left(\frac{v_1}{v_2}\right)^k = p_1\left(\frac{T_2}{T_1}\right)^{\frac{k}{k-1}}$$
$$T_2 = T_1\left(\frac{v_1}{v_2}\right)^{k-1} = T_1\left(\frac{p_2}{p_1}\right)^{\frac{k-1}{k}}$$
$$v_2 = v_1\left(\frac{p_1}{p_2}\right)^{\frac{1}{k}} = v_1\left(\frac{T_1}{T_2}\right)^{\frac{1}{k-1}}$$
$$Q = 0$$
$$u_2 - u_1 = -W = c_v(T_2 - T_1) = \frac{p_2 v_2 - p_1 v_1}{k-1}$$
$$W = u_2 - u_1 = c_v(T_1 - T_2) = \frac{p_1 v_1 - p_2 v_2}{k-1}$$
$$s_2 - s_1 = 0$$
$$h_2 - h_1 = c_p(T_2 - T_1) = \frac{k(p_2 v_2 - p_1 v_1)}{k-1}$$

ENERGY, WORK, AND POWER CONVERSIONS

multiply	by	to get
BTU	3.929 EE–4	hp-hrs
BTU	778.3	ft-lbf
BTU	2.930 EE–4	kw-hrs
BTU	1.0 EE–5	therms
BTU/hr	0.2161	ft-lbf/sec
BTU/hr	3.929 EE–4	hp
BTU/hr	0.2930	watts
ft-lbf	1.285 EE–3	BTU
ft-lbf	3.766 EE–7	kw-hrs
ft-lbf	5.051 EE–7	hp-hrs
ft-lbf/sec	4.6272	BTU/hr
ft-lbf/sec	1.818 EE–3	hp
ft-lbf/sec	1.356 EE–3	kw
hp	2545.0	BTU/hr
hp	550	ft-lbf/sec
hp	0.7457	kw
hp-hr	2545.0	BTU
hp-hr	1.976 EE6	ft-lbf
hp-hr	0.7457	kw-hrs
kw	1.341	hp
kw	3412.9	BTU/hr
kw	737.6	ft-lbf/sec

GENERAL POWER CYCLE

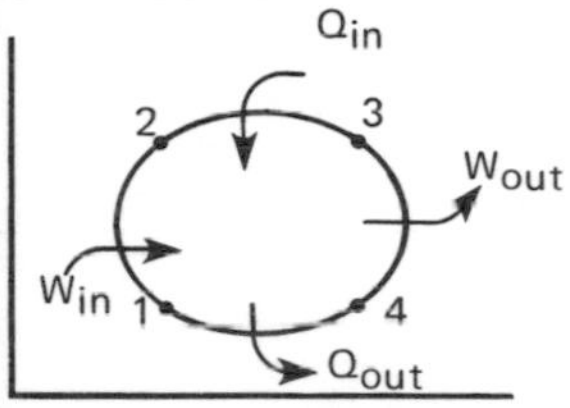

The general power cycle moves *clockwise* on the p-v and T-s diagrams.

1→2: compression
2→3: heat addition
3→4: expansion
4→1: heat rejection

THERMAL EFFICIENCY OF THE ENTIRE CYCLE

$$\eta_{th} = \frac{Q_{in} - Q_{out}}{Q_{in}} = \frac{W_{out} - W_{in}}{Q_{in}}$$

CARNOT CYCLE EFFICIENCY

For the Carnot cycle, the thermal efficiency does not depend on the working fluid. It can be calculated directly from the two temperature extremes.

$$\eta_{th,\text{Carnot}} = \frac{T_{high} - T_{low}}{T_{high}} \quad (T \text{ in } ^\circ R)$$

ISENTROPIC EFFICIENCY OF A PUMP

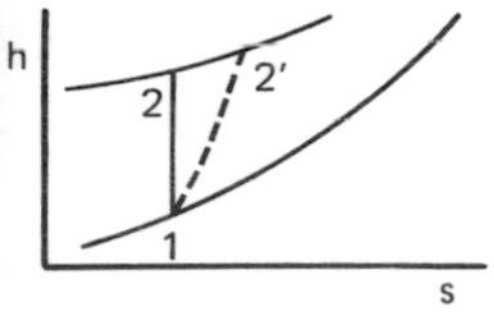

$$\eta = \frac{h_2 - h_1}{h_2' - h_1}$$

$$h_2' = h_1 + \frac{h_2 - h_1}{\eta}$$

ISENTROPIC EFFICIENCY OF A TURBINE

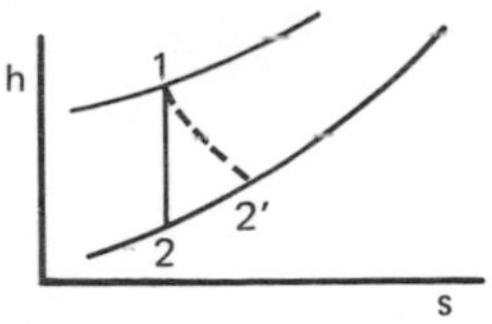

$$\eta = \frac{h_1 - h_2'}{h_1 - h_2}$$

$$h_2' = h_1 - \eta(h_1 - h_2)$$

HEAT OF COMBUSTION

The heat of combustion of a fuel (also known as the *heating value*) is the amount of energy given off when a unit of fuel is burned. Units of *heating value* are BTU/lbm, BTU/gal, and BTU/ft^3, depending on whether the fuel is solid, liquid, or gas. The higher heating value (HHV) includes the heat of vaporization of the water vapor formed.

$$Q_{total} = m(\text{HHV})$$

RATE OF FUEL CONSUMPTION

The rate of fuel consumption in internal combustion engines is known as the *specific fuel consumption* with units of lbm/hp-hr.

$$\text{hourly fuel consumption} = (\text{hp})(\text{SFC})$$

THE PLAN FORMULA

$$\text{hp} = \frac{PLAN}{33{,}000}$$

$$N = \frac{(2n)(\text{no. cylinders})}{\text{no. strokes/cycle}}$$

P is the mean effective pressure in psig

L is the stroke length in feet

A is the bore area in in^2

N is the number of power strokes per minute

n is the engine speed in rpm

HORSEPOWER VERSUS TORQUE

$$(\text{hp})(5252) = (T \text{ in ft-lbf})(\text{rpm})$$

FLOW THROUGH NOZZLES

$$\text{v} = \sqrt{2gJ(h_1 - h_2)} = 223.7\sqrt{h_1 - h_2}$$

RANKINE CYCLE WITH SUPERHEATING

$$W_{turb} = h_d - h_e$$
$$W_{pump} = v_f(p_a - p_f) = h_a - h_f$$
$$Q_{in} = h_d - h_a$$
$$Q_{out} = h_e - h_f$$

The thermal efficiency of the entire cycle is

$$\eta_{th} = \frac{Q_{in} - Q_{out}}{Q_{in}} = \frac{W_{turb} - W_{pump}}{Q_{in}} = \frac{(h_d - h_a) - (h_e - h_f)}{h_d - h_a}$$

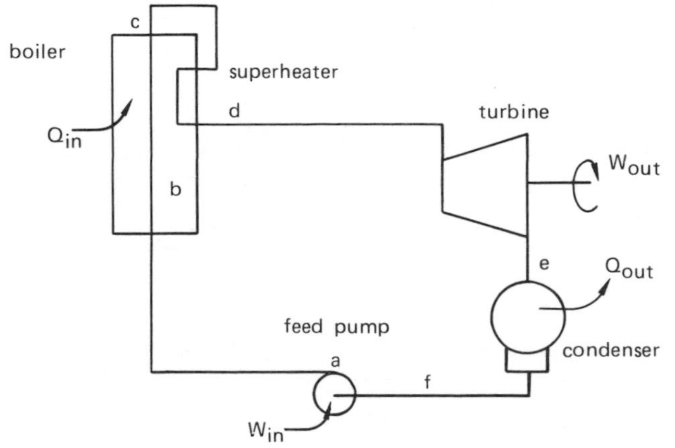

If the Rankine cycle gives efficiencies for the turbine and the pump, calculate all quantities as if those efficiencies were 100%. Then, modify h_e and h_a prior to recalculating the thermal efficiency.

$$h'_e = h_d - \eta_{turb}(h_d - h_e)$$
$$h'_a = h_f + \frac{h_a - h_f}{\eta_{pump}}$$
$$W'_{turb} = h_d - h'_e$$
$$W'_{pump} = h'_a - h_f$$
$$Q'_{in} = h_d - h'_a$$

GENERAL REFRIGERATION CYCLE

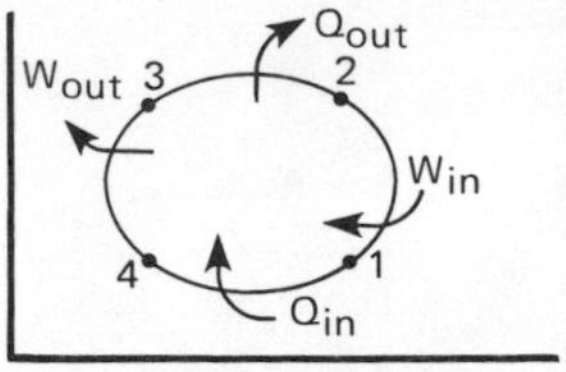

The general refrigeration cycle move *counter-clockwise* on the p-v and T- diagrams.
1→2: compression
2→3: heat rejection
3→4: throttling
4→1: heat addition

REFRIGERATION UNITS

$$1 \text{ ton} = 200 \text{ BTU/min} = 12{,}000 \text{ BTU/hr}$$

HEAT PUMPS

A heat pump operates on a refrigeration cycle using refrig eration equipment. The only difference is the use to which heat pump is put. The purpose of a heat pump is to *warm* the region in which the heat rejection coils are located.

COEFFICIENT OF PERFORMANCE

Efficiencies are not calculated for refrigeration cycles. Rather *coefficients of performance* (COP) are used to compare dif ferent cycles.

$$COP_{refrigerator} = \frac{Q_{absorbed}}{W_{compression}} = \frac{Q_{absorbed}}{Q_{rejected} - Q_{absorbed}}$$

$$COP_{heat\ pump} = \frac{Q_{rejected}}{W_{compression}} = \frac{Q_{absorbed} + W_{compression}}{W_{compression}} = COP_{refrigerator} + 1$$

CARNOT CYCLE COP

$$COP_{refrigerator} = \frac{T_{low}}{T_{high} - T_{low}}$$

$$COP_{heat\ pump} = \frac{T_{high}}{T_{high} - T_{low}} = COP_{refrigerator} + 1$$

ELEMENT AND RADICAL TABLE

name	symbol	atomic weight	valence
acetate	$C_2H_3O_2$	59.0	−1
aluminum	Al	27.0	+3
ammonium	NH_4	18.0	+1
barium	Ba	137.3	+2
boron	B	10.8	+3
borate	BO_3	58.8	−3
bromine	Br	79.9	−1
calcium	Ca	40.1	+2
carbon	C	12.0	+4, −4
carbonate	CO_3	60	−2
chlorate	ClO_3	83.5	−1
chlorine	Cl	35.5	−1
chlorite	ClO_2	67.5	−1
chromate	CrO_4	116.0	−2
chromium	Cr	52.0	+2, +3, +6
copper	Cu	63.6	+1, +2
dichromate	Cr_2O_7	168	−2
fluorine	F	19.0	−1
gold	Au	197.2	+1, +3
hydrogen	H	1.0	+1
hydroxide	OH	17.0	−1
hypochlorite	ClO	51.5	−1
iron	Fe	55.9	+2, +3
lead	Pb	207.2	+2, +4
lithium	Li	6.9	+1
magnesium	Mg	24.3	+2
mercury	Hg	200.6	+1, +2
nickel	Ni	58.7	+2, +3
nitrate	NO_3	62.0	−1
nitrite	NO_2	46.0	−1
nitrogen	N	14.0	−3, +1, +2, +3, +4, +5
oxygen	O	16	−2
permanganate	MnO_4	118.9	−1
phosphate	PO_4	95.0	−3
phosphorous	P	31.0	−3, +3, +5
potassium	K	39.1	+1
silicon	Si	28.1	+4, −4
silver	Ag	107.9	+1
sodium	Na	23.0	+1
sulfate	SO_4	96.1	−2
sulfite	SO_3	80.1	−2
sulfur	S	32.1	−2, +4, +6
tin	Sn	118.7	+2, +4
zinc	Zn	65.4	+2

MOLECULAR WEIGHT OF GASES

Several important gases form diatomic molecules. These are oxygen (O_2), hydrogen (H_2), chlorine (Cl_2), and nitrogen (N_2).

EQUIVALENT WEIGHT

The equivalent weight of an element or radical is its molecular weight divided by its charge when ionized.

MOLES

A mole of a substance is the amount of that substance that has a mass equal to its molecular weight. One gram-mole (gmole) of any gas contains 6.023 EE23 molecules and occupies 22.4 liters at scientific STP. One pound-mole (pmole) of any gas occupies 359.3 ft^3 at scientific STP.

TRIANGLE RULE

The triangle rule is a method of solving weight/proportion problems for binary compounds.

If $n_1A + n_2B \rightarrow A_{n_1}B_{n_2}$

then $\dfrac{w_A}{w_B} = \dfrac{n_1(MW)_A}{n_2(MW)_B}$

STRENGTHS OF SOLUTIONS

The *normality* (N) is the number of gram-equivalent weights of solute per liter of solution.
The *molarity* (M) is the number of gram-moles of solute per liter of solution.
The *molality* (m) is the number of gram-moles of solute per 1000 grams of solvent.

The solution strength may also be expressed in *parts per million* (ppm), which is the number of mass units of solute per million mass units of solution. For water solutions, this is the same as *milligrams per liter* (mg/l).

BOILING POINT/FREEZING POINT CHANGES

The boiling point will rise and the freezing point will drop when a solute is added to a solvent. K_b and K_f depend on the solvent only. For water, $K_b = 0.512$ °C/m and $K_f = 1.86$ °C/m

$$\Delta T_b = (K_b)(\text{solution molality})$$

$$\Delta T_f = (K_f)(\text{solution molality})$$

OXIDATION-REDUCTION REACTIONS

Oxidation is said to occur when an element or compound loses electrons and becomes less negative (more positive). This occurs at the anode.
Reduction is said to occur when an element or compound gains electrons and becomes more negative (less positive). This occurs at the cathode.

pH AND pOH

Let [X] be the ion concentration of ion X in units of moles per liter. Then,

$$\text{pH} = -\log_{10}[\text{H}^+]$$
$$\text{pOH} = -\log_{10}[\text{OH}^-]$$
$$\text{pH} + \text{pOH} = 14$$

The ion concentration may be calculated from

$$[\text{X}] = (\text{fraction ionized})(\text{molarity})$$

ACID/BASE NEUTRALIZATION: TITRATION

(acid volume)(acid normality) =
(base volume)(base normality)

EQUILIBRIUM CONSTANTS

Consider the following reversible reaction:

$$\mathrm{aA} + \mathrm{bB} \longleftrightarrow \mathrm{cC} + \mathrm{dD}$$

The equilibrium constant is

$$K_{eq} = \frac{[\mathrm{C}]^c[\mathrm{D}]^d}{[\mathrm{A}]^a[\mathrm{B}]^b}$$

[X] is the concentration of X.

IONIZATION CONSTANTS

$$K_{ion} = K_{eq}[\mathrm{H_2O}]$$

$$= \frac{(\text{molarity})(\text{fraction ionized})^2}{1 - \text{fraction ionized}}$$

FARADAY'S LAW OF ELECTROLYSIS

The mass of a gas generated by electrolysis is proportional to the amount of electricity used. The name *Faraday* is given to the derived constant which has a value of 96,500 amp-seconds. Then, one Faraday will produce one gram-equivalent weight of a substance through electrolysis.

$$\text{number of grams} = \frac{(I,\ \text{amps})(t,\ \text{seconds})(\text{atomic wt})}{(\text{change in charge})(96,500)}$$

COMPOSITION OF AIR

by weight: 23.15% oxygen, 76.85% nitrogen and others
by volume: 20.9% oxygen, 79.1% nitrogen and others

GALVANIC SERIES

(Anodic to Cathodic)

magnesium alloys
alclad 3S
aluminum alloys
low-carbon steel
cast iron
stainless – No. 410
stainless – No. 430
stainless – No. 404
stainless – No. 316
hastelloy A
lead-tin alloys
brass
copper
bronze
90/10 copper-nickel
70/30 copper-nickel
inconel
silver
stainless steels (passive)
monel
hastelloy C
titanium

PROPERTIES OF COMMON FUELS

	gasoline	octane	propane	ethanol	no. 1 diesel	no. 2 diesel
chemical formula	–	C_8H_{18}	C_3H_8	C_2H_5OH	–	–
molecular weight	≈ 126	114	44	46	≈ 170	≈ 184
carbon % by weight	–	84	82	52	–	–
hydrogen % by weight	–	16	18	13	–	–
oxygen % by weight	–	–	–	35	–	–
heating value						
higher BTU/lbm	20,260	20,590	21,646	12,800	19,240	19,110
lower BTU/lbm	18,900	19,100	19,916	11,500	18,250	18,000
BTU/gal (lower)	116,485	111,824	81,855	76,152	133,332	138,110
latent heat of vaporization						
BTU/lbm	142	141	147	361	115	105
specific gravity	0.739	0.702	0.493	0.794	0.876	0.920
research octane	85–94	100	112	106		–
motor octane	77–86	100	97	89	10–30	
cetane number	10–20	–	–	−20–8	≈ 45	–
stoichiometric mass A/F ratio	14.7	15.1	–	9.0	–	–
distillation temperature (°F)	90–410	–	–	173	340–560	–
flammability limits (volume percent)	1.4–7.6	–	–	4.3–19	–	–

FORCES

A force is a vector quantity. As such, it is completely defined by giving its magnitude, line of application, sense, and point of application. Units of force are pounds in the English system and newtons in the SI system.

RESULTANTS

The resultant of n forces with components $F_{x,i}$ and $F_{y,i}$ has a magnitude of

$$R = \sqrt{(\Sigma F_{x,i})^2 + (\Sigma F_{y,i})^2}$$

The direction is

$$\phi = \arctan\left(\frac{\Sigma F_{y,i}}{\Sigma F_{x,i}}\right)$$

COUPLES

A couple is a moment created by two equal forces acting parallel but with opposite directions. The effect of a couple is to cause rotation. If d is the separation distance, then

$$M = Fd$$

FORCES IN VECTOR FORM

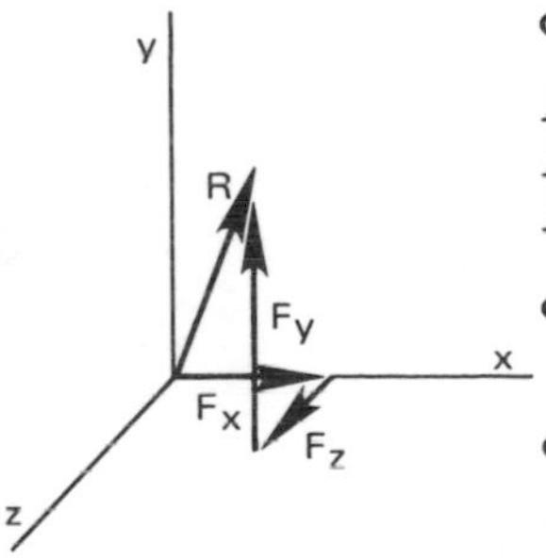

A force, R, may be separated into its components by using the direction cosines.

$$F_x = R\ \cos\theta_x$$
$$F_y = R\ \cos\theta_y$$
$$F_z = R\ \cos\theta_z$$

$$\cos\theta_x = \frac{x}{\sqrt{x^2+y^2+z^2}}$$

$$\cos\theta_y = \frac{y}{\sqrt{x^2+y^2+z^2}}$$

$$\cos\theta_z = \frac{z}{\sqrt{x^2+y^2+z^2}}$$

The resultant may be written in vector form in terms of its components and the unit vectors. (The addition below is vector addition, not algebraic addition.)

$$R = \mathbf{i}F_x + \mathbf{j}F_y + \mathbf{k}F_z$$

COMPONENTS OF INCLINED MEMBERS

A non-trigonometric method can be used to resolve forces in inclined members into their components. This resolution can be accomplished by multiplying the force by the ratio of sides, as determined from the geometry of the inclined member.

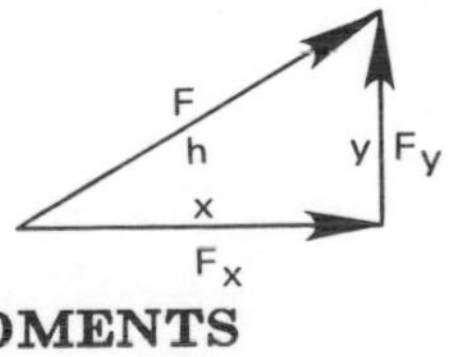

$$F_y = \left(\frac{y}{h}\right)F$$

$$F_x = \left(\frac{x}{h}\right)F$$

MOMENTS

The moment produced by a force acting with a moment arm of length d is

$$M = Fd$$

CONDITIONS FOR EQUILIBRIUM

In general, the conditions for equilibrium are:

$$\Sigma F_x = 0,\ \Sigma F_y = 0,\ \Sigma F_z = 0$$
$$\Sigma M_x = 0,\ \Sigma M_y = 0,\ \Sigma M_z = 0$$

For general coplanar system, the conditions are:

$$\Sigma F_x = 0,\ \Sigma F_y = 0,\ \Sigma M_z = 0$$

For parallel systems, the conditions are:

$$\Sigma F_x = 0,\ \Sigma M_z = 0$$

For concurrent systems, the conditions are:

$$\Sigma F_x = 0,\ \Sigma F_y = 0$$

SIGN CONVENTIONS FOR TRUSS FORCES

For forces in truss members, tension is positive and compression is negative. When freebody diagrams are drawn for truss *joints*, forces leaving the joints place the truss member in tension; forces going into the joints place the member in compression.

DETERMINATE TRUSSES

A truss will be determinate if the number of truss members is equal to

$$\text{no. of truss members} = 2(\text{no. of joints}) - 3$$

If the left-hand side is less than the right-hand side, the truss is not rigid. If the left-hand side is greater than the right-hand side, the truss is indeterminate to the degree of the difference.

CATENARY CABLES

In the figure and equations below, c is the distance from the lowest point on the cable to a reference plane below. The distance c is a constant which must be determined. It does not correspond to any physical dimension.

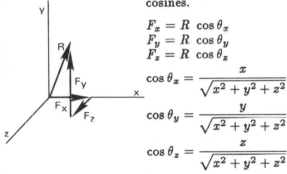

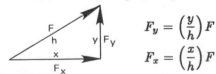

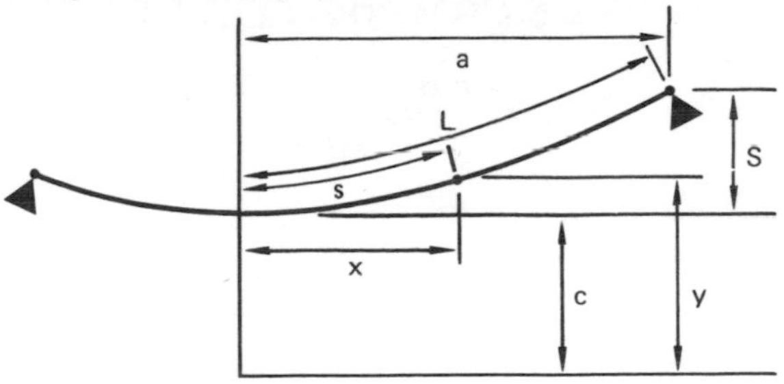

$$y = c\left[\cosh\left(\frac{x}{c}\right)\right]$$

$$s = c\left[\sinh\left(\frac{x}{c}\right)\right]$$

$$y = \sqrt{s^2 + c^2}$$

$$S = c\left[\cosh\left(\frac{a}{c}\right) - 1\right]$$

$$\tan\theta = \frac{s}{c}$$

$H = wc$ (horizontal component of tension)

$F = ws$ (applied vertical load due to cable wt)

$T = wy$ (tangential tension)

FRICTION

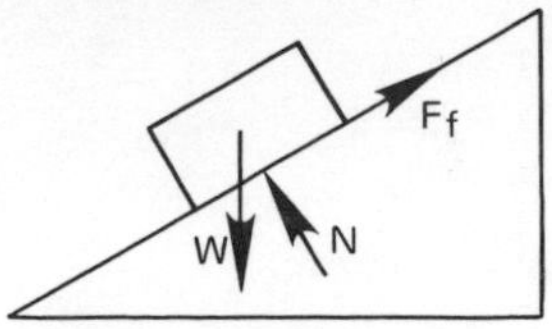

The frictional force depends on the coefficient of friction (μ) and the normal force (N).

$F_f = \mu N$

frictional work $= F_f \times$ distance

CENTROIDS

The centroid of an object is that point at which the object would balance if suspended. In general, the x and y coordinates of the centroidal location can be found from the following relationships:

$$x_c = \frac{1}{A}\int x \, dA \qquad y_c = \frac{1}{A}\int y \, dA$$

The centroidal locations of common shapes are given below:

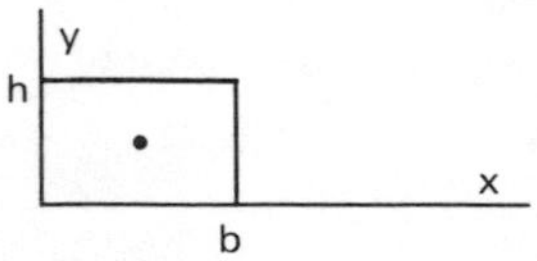

$x_c = \frac{1}{2}b$

$y_c = \frac{1}{2}h$

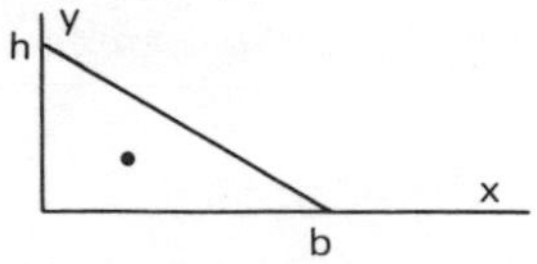

$x_c = \frac{1}{3}b$

$y_c = \frac{1}{3}h$

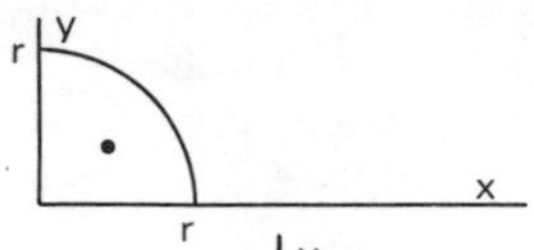

$x_c = \dfrac{4r}{3\pi}$

$y_c = \dfrac{4r}{3\pi}$

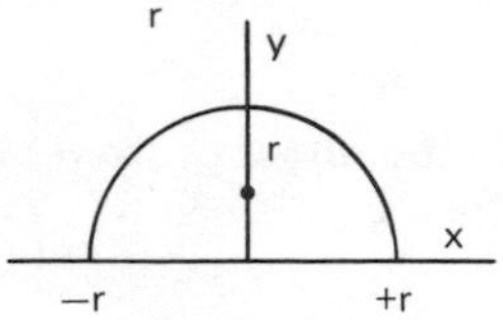

$x_c = 0$

$y_c = \dfrac{4r}{3\pi}$

CENTROIDS OF COMPOSITE (BUILT-UP) SHAPES

$$x_c = \frac{\Sigma A_i x_{c,i}}{\Sigma A_i} \qquad y_c = \frac{\Sigma A_i y_{c,i}}{\Sigma A_i}$$

MOMENTS OF INERTIA

The moment of inertia (second moment of area) has units of $(\text{length})^4$ and can be considered a measure of resistance to bending. I_x and I_y represent the resistance to bending about the x and y axes, respectively. I_x and I_y are not components.

$$I_x = \int y^2 dA \qquad I_y = \int x^2 dA \qquad dA = dx dy$$

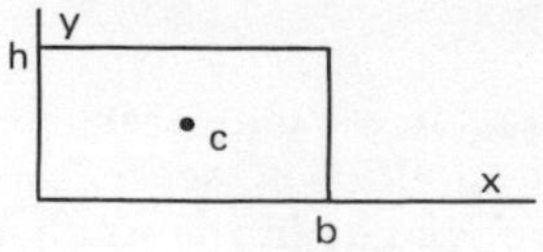

$I_{x,c} = \frac{1}{12}bh^3$

$I_x = \frac{1}{3}bh^3$

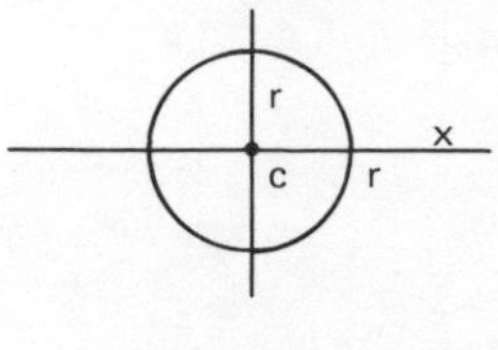

$I_{x,c} = \frac{1}{4}\pi r^4$

RADIUS OF GYRATION

The radius of gyration is the distance from a reference axis at which all of the area can be considered to be concentrated to produce the actual moment of inertia.

$k = \sqrt{\frac{I}{A}}$ or $k = \sqrt{\frac{J}{A}}$ for polar moments of inertia

POLAR MOMENTS OF INERTIA

In general, the polar moment of inertia is

$J = \int r^2 dA = I_x + I_y$

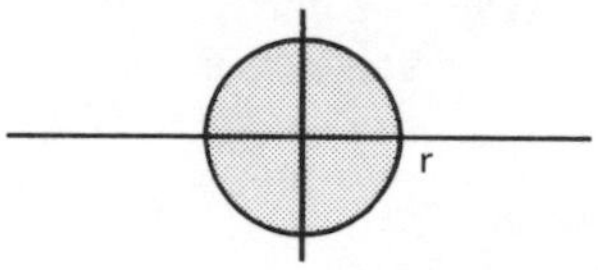

$J = \frac{1}{2}\pi r^4 = \frac{1}{32}\pi d^4$

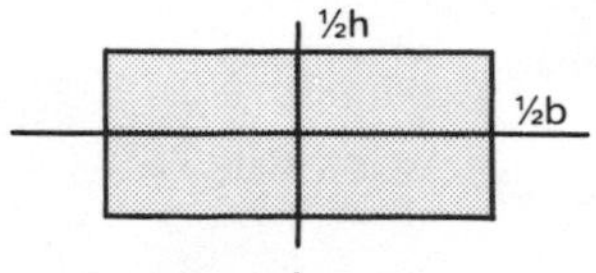

$J = \frac{1}{12}bh(b^2 + h^2)$

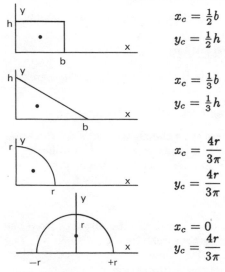

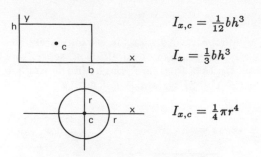

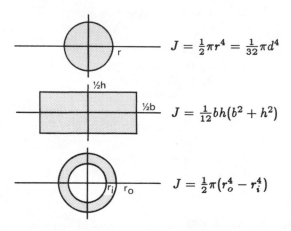

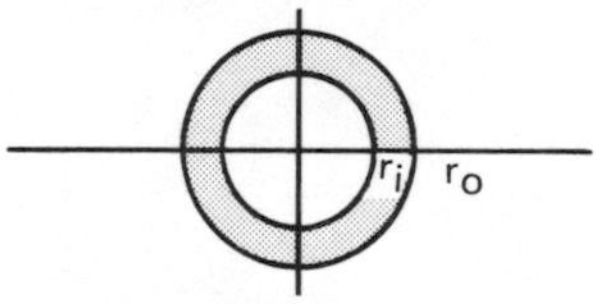

$J = \frac{1}{2}\pi(r_o^4 - r_i^4)$

PARALLEL AXIS THEOREM

If the moment of inertia is known about the centroidal axis, the moment of inertia about a second parallel axis located at distance d away from the centroidal axis is:

$$I_{\text{new}} = I_{\text{centroidal}} + Ad^2 \qquad \text{(area)}$$

$$I_{\text{new}} = I_{\text{centroidal}} + md^2 \qquad \text{(mass)}$$

TENSILE TEST

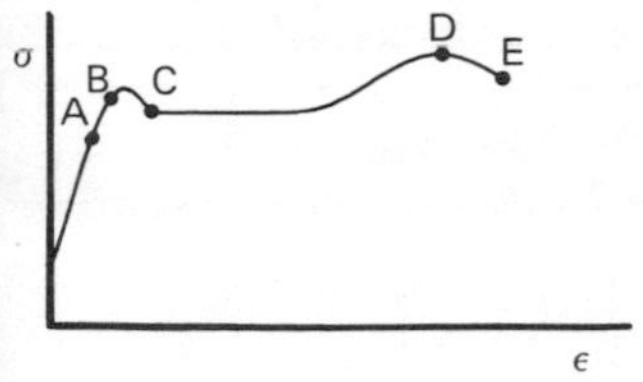

engineering stress: $\sigma = \dfrac{P}{A_o}$

true stress: $\Sigma = \dfrac{P}{A_{\text{inst.}}}$

engineering strain: $\epsilon = \dfrac{\Delta L}{L_o}$

true strain: $E = \ln \dfrac{A_o}{A_{\text{inst.}}}$

point A: proportionality limit—the highest stress for which Hooke's law ($\sigma = E\epsilon$) is valid.

point B: elastic limit—the highest stress for which no permanent deformation occurs.

point C: yield point—the stress at which a sharp drop in load-carrying ability occurs.

point D: ultimate strength—the highest stress which the material can achieve.

point E: fracture strength—the stress at fracture

The *modulus of elasticity (E)* is the slope of the line for stresses up to the proportionality limit.
The *shear modulus (G)* can be calculated from the modulus of elasticity and Poisson's ratio.

$$G = \frac{E}{2(1+\mu)}$$

The *toughness* is the work per unit volume required to cause fracture. It is calculated as the area under the σ-ϵ curve up to the point of fracture. Units of toughness are in-lbf/in^3.
The *ductility* is a measure of the amount of plastic strain at the breaking point. It is calculated as the percent reduction in area at fracture.

$$\text{reduction in area} = \frac{A_o - A_{frac}}{A_o} = \frac{L_{frac} - L_o}{L_o}$$

The *percent elongation at fracture* is calculated from the original length and the fracture length after the sample has "snapped back."

ENDURANCE TEST

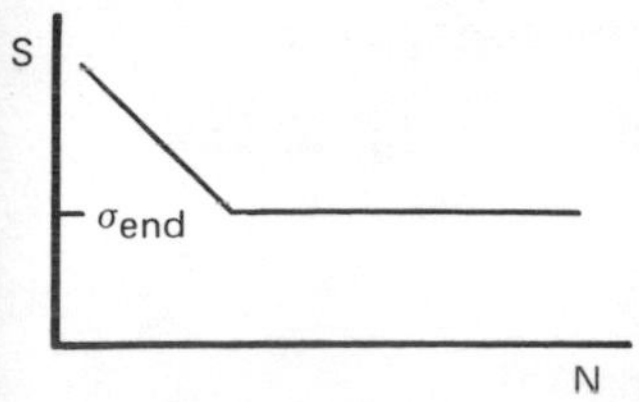

Endurance tests (fatigue tests) apply a cyclical loading of constant maximum amplitude. The plot (usually semi-log or log-log) of the maximum stress and the number of cycles to failure is known as an *S-N plot.*

The *endurance stress (endurance limit* or *fatigue limit)* is the maximum stress which can be repeated indefinitely without causing failure.
The *fatigue life* is the number of cycles required to cause failure for a given stress level.

IMPACT TEST

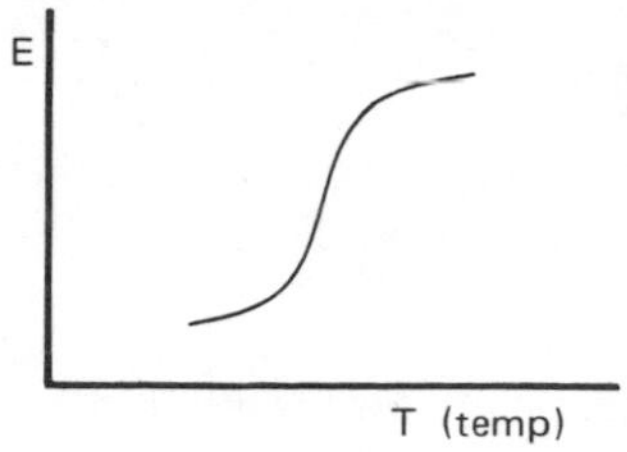

Impact tests determine the amount of energy required to cause failure in standardized test samples. The tests are repeated over a range of temperatures to determine the *transition temperature.* (The transition temperature is approximately 32°F for low-carbon steel).

HARDNESS VERSUS ULTIMATE STRENGTH

$$S_{ut} \approx (500)(\text{BHN}) \qquad \text{for steels}$$

CREEP TEST

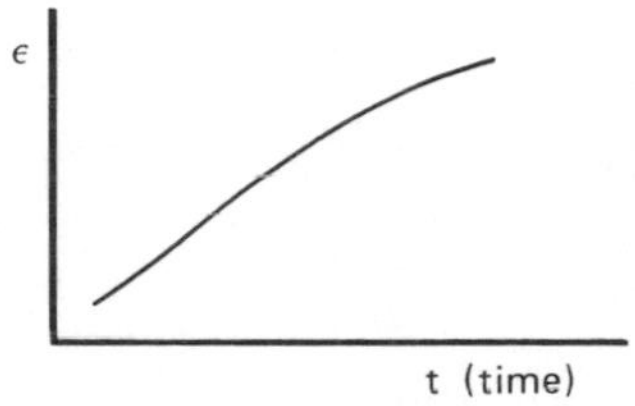

A constant stress less than the yield strength is applied and the elongation versus time is measured.

The *creep rate*$(d\epsilon/dt)$ is very temperature dependent.

The *creep strength* is the stress which results in a given creep rate.
The *rupture strength* is the stress which results in failure after some given amount of time.

CRYSTALLINE STRUCTURE

Crystalline structures are categorized into 14 different point-lattices. Important structures are simple cubic, body-centered cubic (BCC), face-centered cubic (FCC), simple hexagonal, and hexagonal close-packed (HCP).

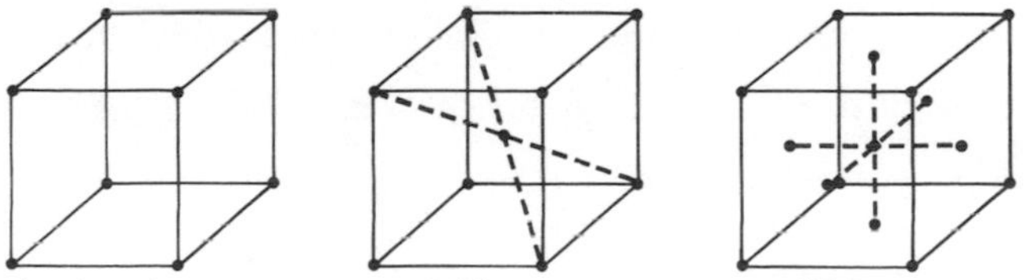
simple cube BCC FCC

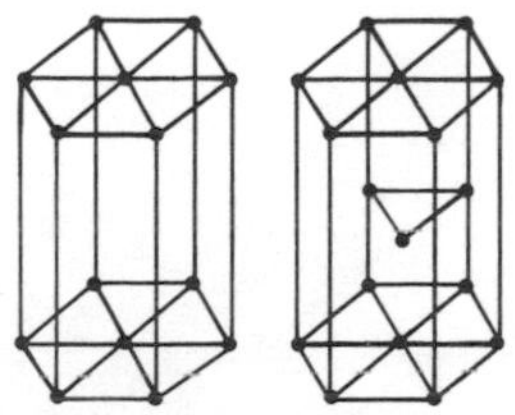
simple hexagonal HCP

Common Metals

The following metals are BBC: chromium, alpha iron, lithium, molybdenum, potassisum, sodium tantalum, and alpha tungsten.
The following metals are FCC: aluminum, copper, gold, lead, nickel, platinum, and silver.
The following metals are HCP: beryllium, cadmium, magnesium, alpha titanium, and zinc.

Number of Atoms in a Cell

simple cubic: 1
BCC: 2
FCC: 4
simple hexagonal: 2
HCP: 6

Packing Factor

The packing factor is the volume of the atoms in a cell (assuming touching, hard spheres) divided by the total cell volume.

simple cubic: 0.52
BCC: 0.68
FCC: 0.74
simple hexagonal: 0.52
HCP: 0.74

Coordination Number

The coordination number is the number of closest neighboring (touching) atoms a given atom has in a lattice.

simple cubic: 6
BCC: 8
FCC: 12
simple hexagonal: 8
HCP: 12

X-RAY DIFFRACTION

If x-rays of wavelength λ are used to find the spacing (d_{hkl}) of crystalline planes, the relationship between the diffraction angle and the nth reinforcement is

$$n\lambda = 2(d_{hkl}) \sin\theta \quad \text{(Bragg's law)}$$

$$\text{Angstrom units:} \quad \dot{A} = \text{EE–8 cm} = \text{EE–10 m}$$

SPACING BETWEEN CRYSTALLINE PLANES

Planar spacing in cubic systems is

$$d_{hkl} = \frac{a}{\sqrt{h^2 + k^2 + l^2}}$$

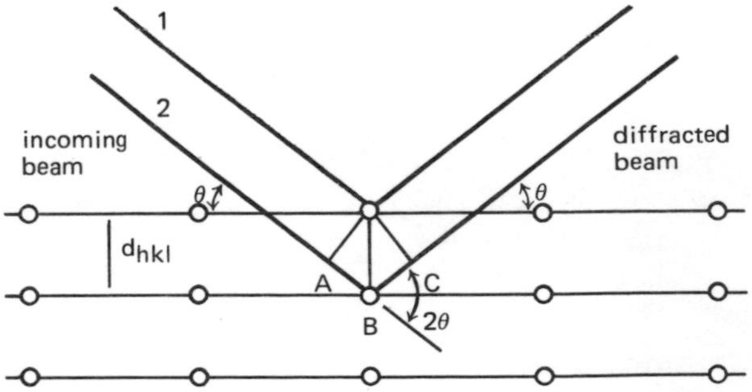

CRYSTALLINE DIRECTIONS

Crystalline directions are specified by a system based essentially on a coordinate system defined by the unit cell. A direction which starts from (0,0,0) and intersects the unit cell at (u,v,w) is specified as [uvw]. No commas are used. Negative numbers are written as positive numbers with overbars.

CRYSTALLINE PLANES: MILLER INDICES

A plane satisfies the equation $\frac{x}{a} + \frac{y}{b} + \frac{z}{c} = 1$ where a, b, and c are the intercepts of the plane on the x, y, and z axes, respectively. If a, b, and c are the intercepts on the axes within a unit cell, the *Miller indices* are written as (hkl) where $h = \frac{1}{a}$, $k = \frac{1}{b}$, and $l = \frac{1}{c}$. No commas are used. Negative numbers are written as positive numbers with overbars.

STEEL ALLOYING INGREDIENTS

(XX is the carbon content, 0.XX%)

alloy number	major alloying elements
10XX	plain carbon steel
11XX	resulfurized plain carbon
13XX	manganese
23XX, 25XX	nickel
31XX,33XX	nickel, chromium
40XX	molybdenum
41XX	chromium, molybdenum
43XX	nickel, chromium, molybdenum
46XX, 48XX	nickel, molybdenum
51XX	chromium
61XX	chromium, vanadium
81XX, 86XX, 87XX	nickel, chromium, molybdenum
92XX	silicon

ALUMINUM ALLOYING INGREDIENTS

alloy number	major alloying ingredient
1XXX	commercially pure aluminum (99 + %)
2XXX	copper
3XXX	manganese
4XXX	silicon
5XXX	magnesium
6XXX	magnesium and silicon
7XXX	zinc
8XXX	other

ALUMINUM TEMPERS

temper	description
T2	annealed (castings only)
T3	solution heat-treated, followed by cold working
T4	solution heat-treated, followed by natural aging
T5	artificial aging
T6	solution heat-treated, followed by artificial aging
T7	solution heat-treated, followed by stabilizing by overaging heat treating
T8	solution heat-treated, followed by cold working and subsequent artificial aging

FICK'S LAWS OF DIFFUSION

$$J = -D\frac{dC}{dx}$$

$$\frac{dC}{dt} = D\frac{d^2C}{dx^2}$$

STRESS

Normal stress: $\sigma = F/A$
Shear stress: $\tau = F/A$

STRAIN (F in y direction)

$\epsilon_y = \Delta L/L_o = F/AE$
$\epsilon_x = \mu\epsilon_y$
$\delta_y = E\epsilon_y$

VALUES OF E (MODULUS OF ELASTICITY) AND G (SHEAR MODULUS)

Steel: $E = 3$ EE7 psi; $G = 1.15$ EE7 psi
Aluminum: $E = 1$ EE7 psi; $G = 3.85$ EE6 psi

The shear modulus, if unknown, can be calculated from the modulus of elasticity and Poisson's ratio.

$$G = \frac{E}{2(1+\mu)}$$

HOOKE'S LAW

For normal stress: $\sigma = E\epsilon$
For shear stress: $\tau = G\epsilon_z$

ELONGATION UNDER NORMAL STRESS

$$\Delta L = L_o\epsilon = \frac{FL_o}{AE}$$

POISSON'S RATIO

μ is the ratio of traverse strain to longitudinal strain.

$$\mu = \frac{\Delta D L_o}{D_o \Delta L}$$

Typical values of μ are 0.3 for steel and 0.33 for aluminum.

THERMAL STRESS AND STRAIN

The elongation of an object when heated is

$$\Delta L = \alpha L_o \Delta T$$

Values of α are 6 EE–6 in/in-°F for steel and 1.3 EE–5 in/in-°F for aluminum.

The thermal stress and strain are:

$$\epsilon_{th} = \alpha\Delta T$$
$$\sigma_{th} = \epsilon_{th}E = \alpha\Delta T E$$

NORMAL STRESS IN A BEAM

Normal stress in a beam is also called "flexure stress" and "bending stress."

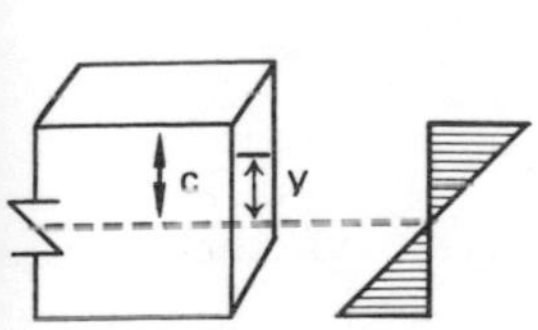

$$\sigma = \frac{My}{I}$$
$$\sigma_{max} = \frac{Mc}{I} = \frac{M}{Z}$$
$$I = \frac{bh^3}{12} \text{ for a rectangle}$$
$$Z = \text{section modulus} = \frac{I}{c}$$

SHEAR STRESS IN A BEAM

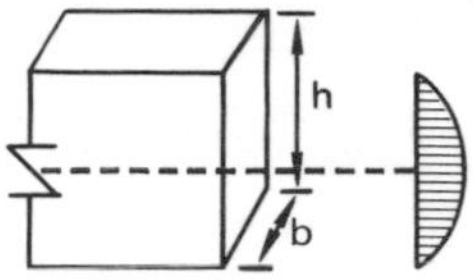

$$\tau = \frac{VQ}{Ib}$$
$$\tau_{max} = \frac{3V}{2A} = \frac{3V}{2bh} \text{ (rectangle)}$$
$$= \frac{4V}{3A} = \frac{4V}{3\pi r^2} \text{ (circular)}$$

V = shear (pounds)

Q = statical moment, also known as "moment of the area"

TORSIONAL STRESS AND STRAIN IN A SHAFT

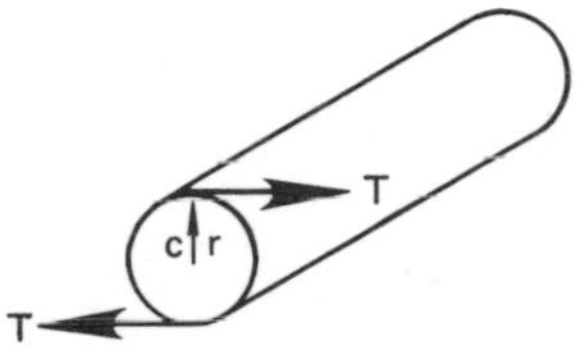

$$\tau = \frac{Tc}{J}$$

J = polar (area) moment of inertia

$= \frac{1}{2}\pi r^4$ (round shafts)

Angle of twist $= \phi = \frac{TL}{GJ}$ (in radians)
The maximum torque that the shaft can carry is

$$T_{max} = \frac{\tau_{max}J}{r}$$

The horsepower transmitted by the shaft is

$$\text{hp} = \frac{2\pi T(\text{rpm})}{33{,}000} \quad (T \text{ in ft-lbf})$$

ECCENTRIC NORMAL STRESS

$$\sigma_{max}, \sigma_{min} = \frac{F}{A} \pm \frac{Mc}{I}$$
$$= \frac{F}{A} \pm \frac{Fec}{I}$$

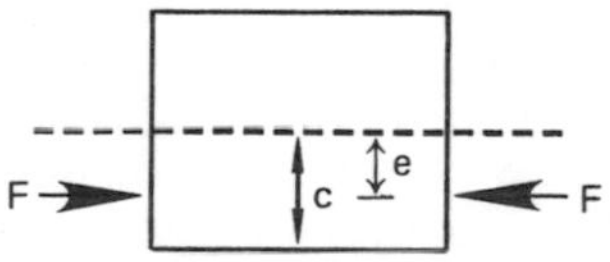

ALLOWABLE STRESS

The allowable stress may be calculated from either the yield stress (typical for ductile materials like steel) or the ultimate strength (typical for brittle materials like cast iron).

$$\sigma_{allowable} = \frac{S_{yield}}{\text{factor of safety}} \quad \text{(ductile)}$$
$$= \frac{S_{ultimate}}{\text{factor of safety}} \quad \text{(brittle)}$$

COMBINED STRESS

The normal and shear stresses on a plane whose normal is inclined an angle θ from the horizontal are

$$\sigma_\theta = \tfrac{1}{2}(\sigma_x + \sigma_y) + \tfrac{1}{2}(\sigma_x - \sigma_y)\cos 2\theta + \tau \sin 2\theta$$

$$\tau_\theta = -\tfrac{1}{2}(\sigma_x - \sigma_y)\sin 2\theta + \tau \cos 2\theta$$

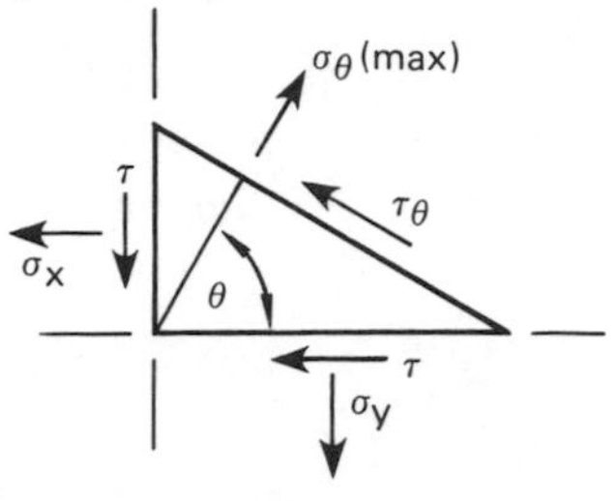

The maximum and minimum values of σ_θ and τ_θ (as θ is varied) are the ***principal stresses.*** These are:

$$\sigma(\text{max, min}) = \tfrac{1}{2}(\sigma_x + \sigma_y) \pm \tau(\text{max})$$

$$\tau(\text{max, min}) = \pm\tfrac{1}{2}\sqrt{(\sigma_x - \sigma_y)^2 + (2\tau)^2}$$

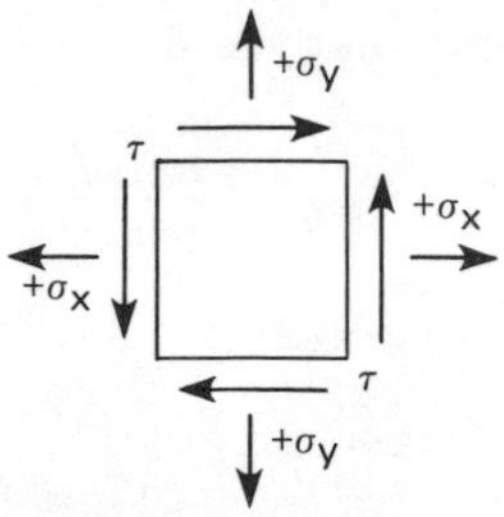

Proper sign convention must be adhered to. Normal tensile stresses are positive; normal compressive stresses are negative. Shear stresses are positive as shown.

MOMENT DIAGRAMS

- Clockwise moments are positive. (Use the left-hand rule.)
- Concentrated loads produce straight inclined lines.
- Uniform loads produce parabolic lines.
- Maximum moment occurs where shear (V) is zero.
- Moment is zero at a free end or hinge.
- Moment at any point is the area under the shear diagram up to that point. That is, $M = \int V\,dx$.

SHEAR DIAGRAMS

- Loads and reactions acting up are positive.
- Shear at any point is the sum of forces up to that point.
- Concentrated loads produce horizontal straight lines.
- Uniform loads produce straight inclined lines.
- Shear at any point is the slope of the moment diagram at that point. That is, $V = \dfrac{dM}{dx}$.

SIMPLE BEAM DEFLECTIONS

(w = load/unit of length)

type of beam	loading	M_{max}	deflection
cantilever	at tip	FL	$\dfrac{FL^3}{3EI}$
cantilever	uniform	$\tfrac{1}{2}wL^2$	$\dfrac{wL^4}{8EI}$
simple	at center	$\tfrac{1}{2}FL$	$\dfrac{FL^3}{48EI}$
simple	uniform	$\dfrac{wL^2}{8}$	$\dfrac{5wL^4}{384EI}$

SLENDER COLUMNS IN COMPRESSION

Euler's formula may be used if the actual stress is kept less than the yield strength. Slender columns fail by buckling.

$$\text{slenderness ratio} = \frac{\text{unbraced length}}{\text{radius of gyration}}$$

$$F_e = \frac{\pi^2 EI}{(CL)^2}$$

$$F_{\text{max}} = \frac{F_e}{\text{F.S.}}$$

$$\sigma_e = \frac{F_e}{A}$$

F.S. = factor of safety

$C = 1$ for both ends pinned (translation fixed, but rotation free)

$C = 0.5$ for both ends fixed against translation and rotation (built-in)

$C = 0.707$ one end pinned and the other end built-in

$C = 2$ for one end fixed and the other end completely free

THIN-WALL TANKS (PIPE STRESS)

A tank has thin walls if the wall thickness is less than 1/1 of the tank diameter.

$$\sigma_{\text{hoop}} = \frac{pr}{t}$$

r = tank radius

t = wall thickness

p = gage pressure inside tank

$$\sigma_{\text{longitudinal}} = \frac{pr}{2t}$$

For spherical tanks or spherical ends on a cylinder tank, use the longitudinal stress formula.

SPRINGS

The force required to elongate a spring with a spring constant of k is

$$F = k\Delta L$$

$$W = \tfrac{1}{2}kx^2$$

$$k = k_1 + k_2 + k_3 + \ldots \quad \text{(parallel springs)}$$

$$\frac{1}{k} = \frac{1}{k_1} + \frac{1}{k_2} + \frac{1}{k_3} + \cdots \quad \text{(series springs)}$$

DISTANCE, VELOCITY, AND ACCELERATION

$$a = \frac{dv}{dt} = \frac{d^2s}{dt^2}$$
$$v = \frac{ds}{dt} = \int a\, dt$$
$$s = \int v\, dt = \int\int a\, dt$$

ROTATIONAL MOTION

Analogous variables exist for rotational motion.

α : rotational acceleration
ω : rotational speed
θ : rotational position

$$\alpha = \frac{d\omega}{dt} = \frac{d^2\theta}{dt^2}$$
$$\omega = \frac{d\theta}{dt} = \int \alpha\, dt$$
$$\theta = \int \omega\, dt = \int\int \alpha\, dt$$

INSTANTANEOUS RELATIONSHIP BETWEEN ROTATIONAL AND LINEAR MOTION

$$\omega = 2\pi\left(\frac{\text{rpm}}{60}\right)$$

$$s = r\theta \qquad v = r\omega \qquad a = r\alpha$$

UNIFORM ACCELERATION FORMULAS

to find	given these	use this formula
t	a, v_o, v	$t = \frac{v - v_o}{a}$
t	a, v_o, s	$t = \frac{\sqrt{2as + v_o^2} - v_o}{a}$
t	v_o, v, s	$t = \frac{2s}{v_o + v}$
a	t, v_o, v	$a = \frac{v - v_o}{t}$
a	t, v_o, s	$a = \frac{2s - 2v_o t}{t^2}$
a	v_o, v, s	$a = \frac{v^2 - v_o^2}{2s}$
v_o	t, a, v	$v_o = v - at$
v_o	t, a, s	$v_o = \frac{s}{t} - \frac{1}{2}at$
v_o	a, v, s	$v_o = \sqrt{v^2 - 2as}$
v	t, a, v_o	$v = v_o + at$
v	a, v_o, s	$v = \sqrt{v_o^2 + 2as}$
s	t, a, v_o	$s = v_o t + \frac{1}{2}at^2$
s	a, v_o, v	$s = \frac{v^2 - v_o^2}{2a}$
s	t, v_o, v	$s = \frac{1}{2}t(v_o + v)$

ACCELERATION DUE TO GRAVITY

Falling body problems may be solved with the above uniform acceleration formulas by substituting

$$a = g = -32.2 \text{ ft/sec}^2$$

PROJECTILE MOTION

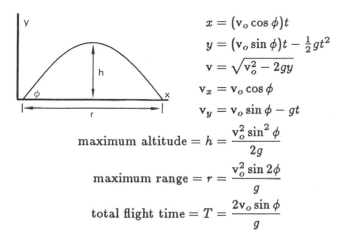

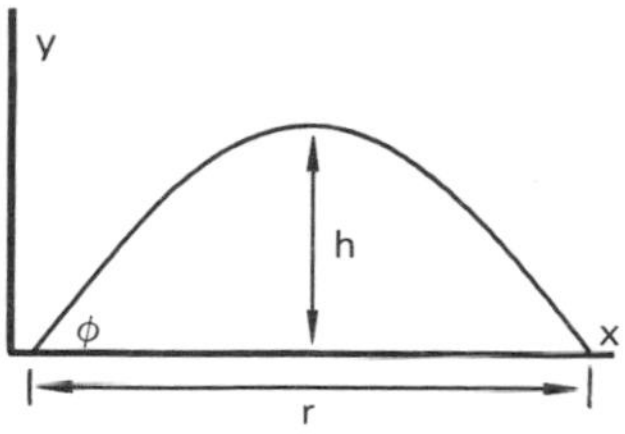

$$x = (v_o \cos\phi)t$$
$$y = (v_o \sin\phi)t - \tfrac{1}{2}gt^2$$
$$v = \sqrt{v_o^2 - 2gy}$$
$$v_x = v_o \cos\phi$$
$$v_y = v_o \sin\phi - gt$$
$$\text{maximum altitude} = h = \frac{v_o^2 \sin^2\phi}{2g}$$
$$\text{maximum range} = r = \frac{v_o^2 \sin 2\phi}{g}$$
$$\text{total flight time} = T = \frac{2v_o \sin\phi}{g}$$

The above formulas neglect air drag. Range (r) is maximum when $\phi = 45°$.

WORK AND ENERGY (m in lbm)

Work is an energy transfer that occurs when a force (or torque, T) is moved through a distance:

$$W = \int F \cdot ds \qquad \text{(linear)}$$
$$W = \int T \cdot d\theta \qquad \text{(rotational)}$$

Potential energy is the energy that an object possesses by virtue of its position in a gravitational field:

$$E_p = \frac{mgh}{g_c} \text{ (in ft-lbf)}$$

For a spring (with rate constant k) compressed an amount x,

$$E_p = \tfrac{1}{2}kx^2$$

Kinetic energy is the energy that an object possesses by virtue of its velocity:

$$E_k = \frac{mv^2}{2g_c} \qquad \text{(linear)}$$
$$E_k = \frac{I\omega^2}{2g_c} \qquad \text{(rotational)}$$

THE WORK-ENERGY PRINCIPLE

In the absence of thermal changes, the work done on (or by) a system is equal to its change in energy.

$$W = \Delta E_p + \Delta E_k$$

POWER

$$P = \frac{W}{t} = F\text{v} = T\omega$$

CENTRIFUGAL FORCE

$$F_c = \frac{m\text{v}^2}{g_c r}$$

$$a_n = \frac{\text{v}_t^2}{r} = r\omega^2 = \text{v}_t\omega$$

HIGHWAY BANKING

The banking angle (superelevation) needed on a circular curve of radius r is

$$\phi = \arctan\frac{\text{v}^2}{gr}$$

HORSEPOWER REQUIRED TO MAINTAIN VELOCITY

For translation, $\text{hp} = \frac{F\text{v}}{550}$ (v in fps)

For rotation, $\text{hp} = \frac{2\pi Tn}{33,000}$ (T in ft-lbf, n in rpm)

NEWTON'S LAWS

The first law: The velocity (momentum) of an object will not change unless it is acted upon by a force.
The second law: $F = ma$ (consistent units)
The third law: For every action there is an equal but opposite reaction.
Law of Universal Gravitation: The gravitational force between two objects with masses m_1 and m_2 is

$$F = \frac{Gm_1m_2}{d^2} \quad \text{(consistent units)}$$

NEWTON'S SECOND LAW FOR ROTATIONAL SYSTEMS

$$T = I\alpha$$

MOMENT OF INERTIA FOR ROTATIONAL SYSTEMS

In general, $I = \int r^2 dm$

For a solid cylinder, $I = \frac{mr^2}{2g_c}$

For a hollow cylinder, $I = \frac{m(r_i^2 + r_o^2)}{2g_c}$

PARALLEL AXIS THEOREM

$$I_{\text{new}} = I_{\text{centroidal}} + \frac{md^2}{g_c}$$

RADIUS OF GYRATION

$$k = \sqrt{\frac{Ig_c}{m}}$$

IMPULSE-MOMENTUM PRINCIPLE

For translational motion, $F\Delta t = \frac{m}{g_c}\Delta\text{v}$

For rotational motion, $T\Delta t = I\Delta\omega$

COLLISIONS

The coefficient of restitution (e) is 1 if the collision is perfectly elastic. e is 0 if the collision is completely inelastic (the bodies stick together).

$$e = -\frac{\text{v}_2' - \text{v}_1'}{\text{v}_2 - \text{v}_1}$$

Momentum is conserved in all collisions, although kinetic energy may not be ($e < 1$). The conservation of momentum equation is:

$$m_1\text{v}_1' + m_2\text{v}_2' = m_1\text{v}_1 + m_2\text{v}_2$$

SIMPLE HARMONIC MOTION

$$\omega = 2\pi f = \frac{2\pi}{T}$$

$$f = \frac{\omega}{2\pi} = \frac{1}{T}$$

$$T = \frac{1}{f} + \frac{2\pi}{\omega}$$

For a simple spring-and-mass oscillator,

$$f = \frac{1}{2\pi}\sqrt{\frac{kg_c}{m}}$$

$$T = 2\pi\sqrt{\frac{m}{kg_c}}$$

For a simple pendulum consisting of a mass (m) at the end of a massless cord of length L,

$$f = \frac{1}{2\pi}\sqrt{\frac{g}{L}}$$

$$T = 2\pi\sqrt{\frac{L}{g}}$$

SYSTEMS OF SPRINGS

For springs in parallel,

$$k_{\text{equivalent}} = \Sigma k_i$$

For springs in series,

$$\frac{1}{k_{\text{equivalent}}} = \sum\left(\frac{1}{k_i}\right)$$

For two springs in series,

$$k_{\text{equivalent}} = \frac{k_1k_2}{k_1 + k_2}$$

PREFIXES

pico (p) = EE–12 micro (μ) = EE–6 kilo (k) = EE3
nano (n) = EE–9 milli (m) = EE–3 meg (M) = EE6

ENERGY CONVERSIONS

1 joule = 1 N-m = EE7 ergs = 1 watt-sec

POWER CONVERSIONS

1 watt = 1 joule/sec
746 watts = 1 horsepower

CHARGE

The charge on an electron is 1 electrostatic unit (ESU), equal to 1.6 EE–19 coulombs. 1 coulomb is equal to 6.24 EE18 ESU.

ELECTROSTATICS

flux: ϕ (in lines)

flux density: $D = \dfrac{\phi}{A}$

For a point charge, $D = \dfrac{q}{4\pi r^2}$

electric field: $E = \dfrac{D}{\epsilon}$

$\epsilon = \epsilon_v \epsilon_r$
= permittivity of the medium

$\epsilon_v = 8.85 \text{ EE–12} \dfrac{\text{coulomb}^2}{\text{newton-meter}^2}$

ϵ_r = dielectric constant (1 for air)

For a point charge, $E = \dfrac{q}{4\pi\epsilon r^2}$

Coulomb's law: $F = q_2 E = \dfrac{q_1 q_2}{4\pi\epsilon r^2}$

work: The work required to change the separation of two charges from r_1 to r_2 is:

$$W = \frac{-q_1 q_2}{4\pi\epsilon}\left(\frac{1}{r_2} - \frac{1}{r_1}\right)$$

system potential energy: $E_p = \dfrac{-q_1 q_2}{4\pi\epsilon r}$

potential: $V = \dfrac{E_p}{q}$

UNIFORM ELECTRIC FIELDS

If a voltage V appears across two plates separated by a distance r (in meters), then the electric field is $E = \frac{V}{r}$.

The force that a charged particle feels is $F = Eq$.

The work done in moving a particle a distance d in the uniform field is $W = Fd = Eqd = \dfrac{Vqd}{r}$.

MAGNETISM

flux: ϕ (in unit-poles) $= BA = M$

flux density: $B = \dfrac{\phi}{A}$

For a pole of strength M, $B = \dfrac{M}{4\pi r^2}$

magnetic field: $H = \dfrac{B}{u}$

$u = u_v u_r$
= permeability of the medium

$u_v = 4\pi \text{ EE–7} \dfrac{(\text{unit-poles})^2}{\text{newton-meter}^2}$

force: $F = M_2 H = \dfrac{M_1 M_2}{4\pi\mu r^2}$

MAGNETIC FIELD AROUND A WIRE CARRYING CURRENT

$H = \dfrac{B}{u} = \dfrac{I}{2\pi r}$ (r is distance from wire center)

$F = IBL$

FARADAY'S LAW

The induced voltage in n coils is

$$V = n\frac{d\phi}{dt} = \frac{d\lambda}{dt} \quad (\lambda \text{ is the "flux linkage"})$$

OHM'S LAW

$V = IR$

RESISTORS IN SERIES AND PARALLEL

series: $R_c = \Sigma R_i$

parallel: $\dfrac{1}{R_c} = \sum \dfrac{1}{R_i}$

For 2 resistors in parallel: $R_c = \dfrac{R_1 R_2}{R_1 + R_2}$

RESISTIVITY

The resistance of a material whose resistivity is ρ and whose length and area are L and A, respectively, is

$$R = \frac{\rho L}{A}$$

A is usually measured in circular mils. A circular mil is the area of a 0.001″ diameter circle. Therefore,

$$A = \left(\frac{D \text{ in inches}}{0.001}\right)^2$$

Resistance varies with temperature according to

$$R = R_o(1 + \alpha\Delta T)$$
$$\rho = \rho_o(1 + \alpha\Delta T)$$

POWER IN A RESISTIVE CIRCUIT

$$P = IV = I^2 R = \frac{V^2}{R} \qquad \text{(watts)}$$

DECIBELS

$$dB = 10 \log_{10}\frac{P_2}{P_1}$$
$$20 \log_{10}\frac{V_2}{V_1} \quad \text{if impedances are the same}$$

KIRCHOFF'S LAWS

Kirchoff's current law: $\Sigma I_{\text{in}} = \Sigma I_{\text{out}}$ at a junction
Kirchoff's voltage law: $\Sigma V = \Sigma(IR)$ around a loop

DELTA-WYE CONVERSIONS

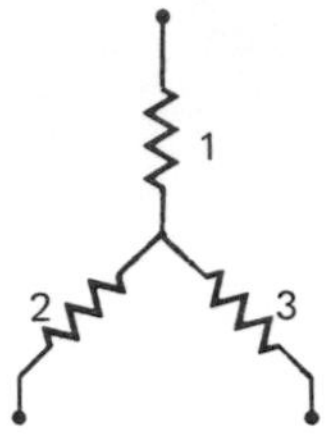

Let $R = R_a + R_b + R_c$

Then, $R_1 = \dfrac{R_a R_c}{R}$

$R_2 = \dfrac{R_a R_b}{R}$

$R_3 = \dfrac{R_b R_c}{R}$

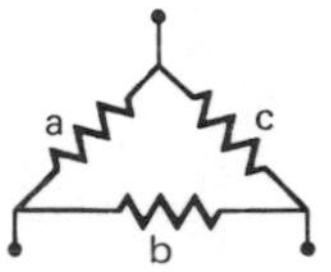

Let $R = R_1R_2 + R_1R_3 + R_2R_3$

Then, $R_a = \dfrac{R}{R_3}$

$R_b = \dfrac{R}{R_1}$

$R_c = \dfrac{R}{R_2}$

WHEATSTONE BRIDGE

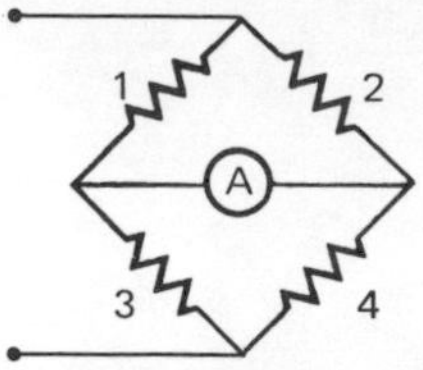

If the ammeter reads zero, then
$I_2 = I_4$
$I_1 = I_3$
$V_1 + V_3 = V_2 + V_4$

THEVENIN EQUIVALENT

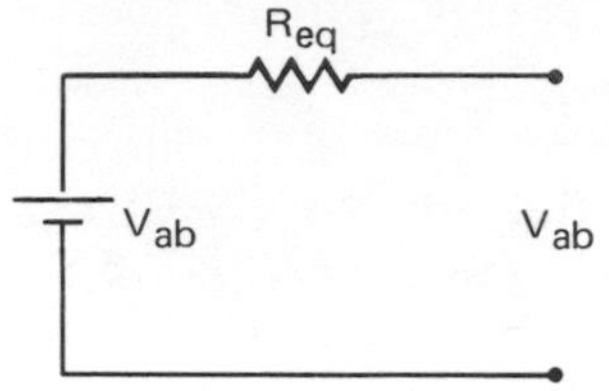

$$R_{eq} = \frac{V_{ab}}{I}$$

NORTON EQUIVALENT

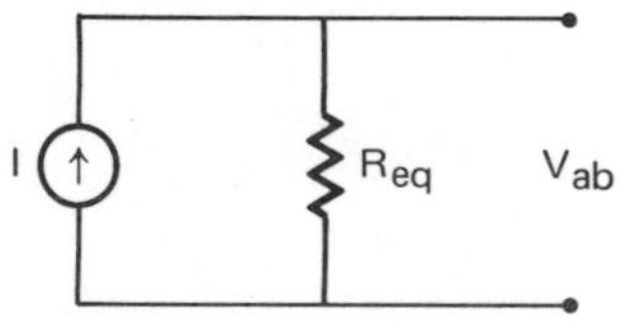

$$R_{eq} = \frac{V_{ab}}{I}$$

D.C. TRANSIENT SOLUTIONS

type of circuit	charging	discharging
series RC: time constant = RC seconds 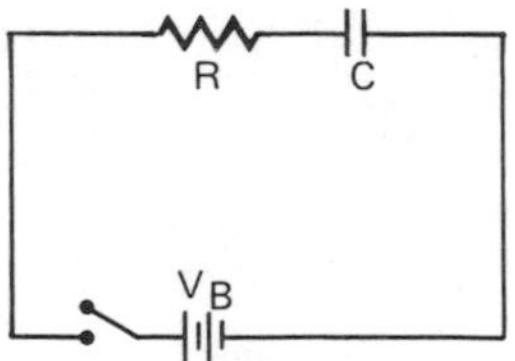 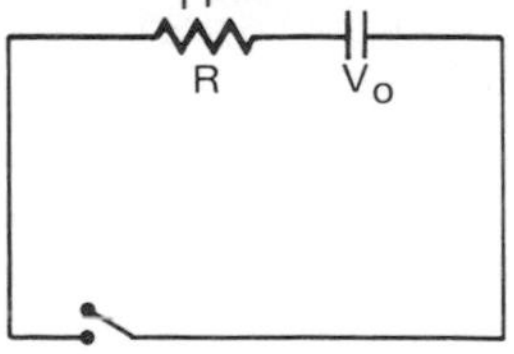 	$V_R(t) = V_B e^{-t/RC}$ $V_C(t) = V_B\left(1 - e^{-t/RC}\right)$ $I(t) = \left(\frac{V_B}{R}\right)e^{-t/RC}$ $V_B = V_R + V_C$ $V_B = I(t)R + \left(\frac{1}{C}\right)I(t)dt$ $V_B = R\left(\frac{dq}{dt}\right) + \frac{q}{C}$ $q_C(t) = CV_B\left(1 - e^{-t/RC}\right)$	$V_R(t) = -V_o e^{-t/RC}$ $V_C(t) = V_o e^{-t/RC}$ $I(t) = \left(\frac{-V_o}{R}\right)e^{-t/RC}$ $0 = V_R + V_C$ $q_C(t) = CV_o\left(e^{-t/RC}\right)$
series RL: time constant = L/R seconds 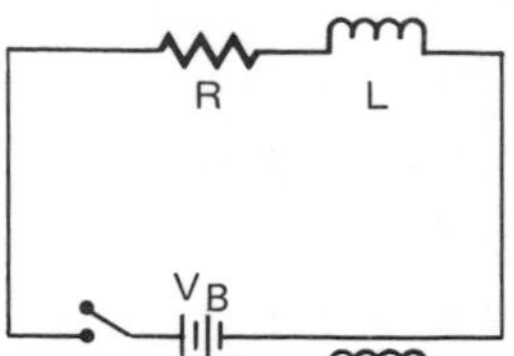 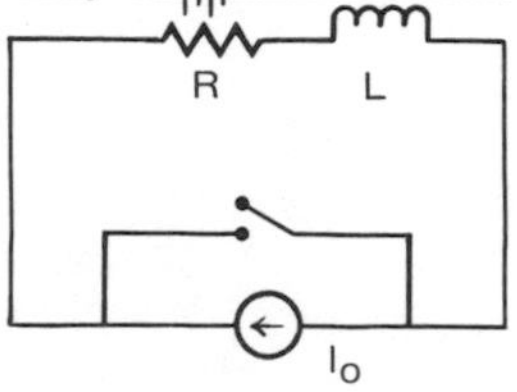 	$V_R(t) = V_B\left(1 - e^{-Rt/L}\right)$ $V_L(t) = V_B e^{-Rt/L}$ $I(t) = \left(\frac{V_B}{R}\right)\left(1 - e^{-Rt/L}\right)$ $V_B = V_R + V_L$ $V_B = RI(t) + L\left(\frac{dI(t)}{dt}\right)$	$V_R(t) = I_oR\left(e^{-Rt/L}\right)$ $V_L(t) = -I_oR\left(e^{-Rt/L}\right)$ $I(t) = I_o e^{-Rt/L}$ $0 = V_R + V_L$

TIME, PERIOD, AND FREQUENCY

$$\omega = 2\pi f = \frac{2\pi}{T}$$

$$f = \frac{\omega}{2\pi} = \frac{1}{T}$$

$$T = \frac{2\pi}{\omega} = \frac{1}{f}$$

GENERATED FREQUENCY

$$f = \frac{(\text{rpm})(\text{number of poles})}{120}$$

AVERAGE VALUES

For a repeating waveform with period T, the average value is

$$S_{\text{ave}} = \frac{1}{T}\int_0^T S(t)\,dt$$

The average value of a rectified sinusoid is

$$\frac{2S_{\max}}{\pi}$$

EFFECTIVE VALUES

For a repeating waveform with period T, the effective value (also known as "root-mean-square" value) is

$$S_{\text{eff}} = \sqrt{\frac{1}{T}\int_0^T S^2(t)\,dt}$$

The effective value of a sinusoid is

$$\frac{S_{\max}}{\sqrt{2}} = 0.707S_{\max}$$

FACTORS

The form factor is $\frac{S_{\text{eff}}}{S_{\text{ave}}}$. The form factor for a rectified sinusoid is 1.11. The crest factor is $\frac{S_{\max}}{S_{\text{eff}}}$. The crest factor for a sinusoid is 1.41.

SINE-COSINE CONVERSIONS

$$\cos(\omega t) = \sin(\omega t + \tfrac{1}{2}\pi) = -\sin(\omega t - \tfrac{1}{2}\pi)$$

$$\sin(\omega t) = \cos(\omega t - \tfrac{1}{2}\pi) = -\cos(\omega t + \tfrac{1}{2}\pi)$$

RESISTORS

governing equation: $V(t) = I(t)R$
power: $P(t) = \frac{1}{2}I_{\max}V_{\max} - \frac{1}{2}I_{\max}V_{\max}\cos 2\omega t$
average power: $P_{\text{ave}} = \frac{1}{2}I_{\max}V_{\max} = I_{\text{eff}}V_{\text{eff}}$

CAPACITORS (ICE – Leading)

governing equations: $q(t) = CV(t)$

$$V(t) = \left(\frac{1}{C}\right)\int I(t)\,dt$$

$$I = \frac{dq(t)}{dt}$$

capacitance: $C = \frac{\epsilon A}{r}$ where

ϵ_v = 8.85 EE–12 farads/meter
A = plate area in square meters
r = plate separation in meters

reactance: $X_C = \frac{1}{\omega C} = \frac{1}{2\pi f C}$
angle: $\angle -90°$
power: $P(t) = \frac{1}{2}I_{\max}V_{\max}\sin 2\omega t$
average power: $P_{\text{ave}} = 0$
capacitors in series: $\frac{1}{C} = \sum \frac{1}{C_1}$

$$V_{C_1} = \frac{C_2}{C_1 + C_2} \times V_{\text{total}}$$

capacitors in parallel: $C = \sum C_i$

INDUCTORS (ELI – Lagging)

governing equation: $V(t) = L\frac{dI(t)}{dt}$
reactance: $X_L = \omega L = 2\pi f L$
angle: $\angle 90°$
power: $P(t) = -\frac{1}{2}I_{\max}V_{\max}\cos 2\omega t$
average power: $P_{\text{ave}} = 0$
inductors in series: $L = \sum L_i$
inductors in parallel: $\frac{1}{L} = \sum \frac{1}{L_i}$

ADMITTANCE

$$Y = \frac{1}{Z}$$

RESONANCE

At resonance, the frequency is

$$\omega^* = \frac{1}{\sqrt{LC}} = 2\pi f^* \quad \text{(radians/sec)}$$

Series Resonance

$Z = R$
Z is minimum
I is maximum
P is maximum
Q = quality factor $= \omega^*\frac{L}{R} = \frac{1}{\omega^* CR}$

$$\text{bandwidth} = \frac{\omega^*}{Q} = \frac{2\pi f^*}{Q}$$

Parallel Resonance

$Z = R$
Z is maximum
I is minimum
P is minimum

$Q = \text{quality factor} = \omega^* CR = \dfrac{R}{\omega^* L}$

$\text{bandwidth} = \dfrac{\omega^*}{Q} = \dfrac{2\pi f^*}{Q}$

SIMPLE TRANSFORMERS

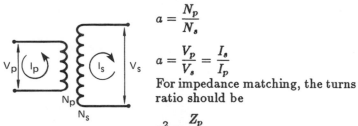

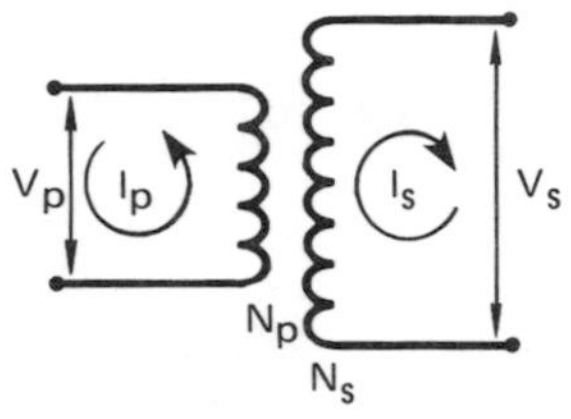

$$a = \frac{N_p}{N_s}$$

$$a = \frac{V_p}{V_s} = \frac{I_s}{I_p}$$

For impedance matching, the turns ratio should be

$$a^2 = \frac{Z_p}{Z_s}$$

COMPLEX POWER

1 hp = 0.7457 kw

$P_{\text{real}} = \frac{1}{2} I_{R,\text{max}} V_{R,\text{max}} = I_{R,\text{eff}} V_{R,\text{eff}} = P_{\text{apparent}} \cos\phi$

$P_{\text{reactive}} = \frac{1}{2} I_{X,\text{max}} V_{X,\text{max}} = I_{X,\text{eff}} V_{X,\text{eff}}$
$= P_{\text{apparent}} \sin\phi$

$P_{\text{apparent}} = \sqrt{P_{\text{real}}^2 + P_{\text{reactive}}^2} = \frac{1}{2} I_{\text{line,max}} V_{\text{line,max}}$

power factor $= \cos\phi$
reactive factor $= \sin\phi$

POWER FACTOR CORRECTION

units: kw for real power
kva for apparent power
kvar for imaginary power

The required change in reactive power to change the power angle from ϕ_1 to ϕ_2 is

$$\Delta P_{\text{reactive}} = P_{\text{real}}(\tan\phi_1 - \tan\phi_2)$$

The capacitance (in farads) required to change the reactive power by the above amount is given below. V_{line} is the maximum value of the sinusoid. (If V_{line} is an rms value, then divide by 2.0.)

$$C = \frac{\Delta P_{\text{reactive}}}{\pi f V_{\text{line}}^2}$$

SYNCHRONOUS MACHINES

$$\text{synchronous speed} = \frac{120 f}{\text{number of poles}}$$

INDUCTION MACHINES

The stator field rotates at synchronous speed (n_s). If the rotor rotates at n_r, the slip is $\dfrac{n_s - n_r}{n_s}$.

3-PHASE LOADS

Delta-Wired Loads

$$V_{\text{line}} = V_{\text{phase}}$$
$$I_{\text{line}} = \sqrt{3}\, I_{\text{phase}}$$
$$P_{\text{total}} = 3P_{\text{phase}} = 3V_{\text{phase}} I_{\text{phase}} \cos\phi$$
$$= \sqrt{3}\, V_{\text{line}} I_{\text{line}} \cos\phi$$

Wye-Wired Loads

$$V_{\text{line}} = \sqrt{3}\, V_{\text{phase}}$$
$$I_{\text{line}} = I_{\text{phase}}$$
$$P_{\text{total}} = 3P_{\text{phase}} = 3V_{\text{phase}} I_{\text{phase}} \cos\phi$$
$$= \sqrt{3}\, V_{\text{line}} I_{\text{line}} \cos\phi$$

Equivalent Loads

A balanced Δ system may be replaced by a balanced Y system if $Z_Y = \frac{1}{3} Z_\Delta$.

BOHR'S ATOM

first postulate: Electrons circle the nucleus in stable, non-radiating orbits.
second postulate: Energy is quantized.
third postulate: Classical mechanics doesn't hold for electrons between orbits.
fourth postulate: Angular momentum is quantized.

The second postulate can be written as

$$E_2 - E_1 = hf$$

h is Planck's constant, which has a value of 6.626 EE–34 J-sec, or 4.136 EE–15 eV-sec.

The fourth postulate can be written as

$$m\text{v}r = \frac{nh}{2\pi}$$

n is the principle quantum number, having a value of 1 for the lowest electron orbit.

SERIES OF HYDROGEN SPECTRA

series name	series type
Lyman	ultraviolet
Balmer	visible
Paschen	infrared
Brackett	infrared
Pfund	infrared

PAULI EXCLUSION PRINCIPLE

No two electrons may have the same set of quantum numbers. Stated another way, no two electrons may occupy the same orbital and both have the same spin.

NUMBER OF ELECTRONS IN A SHELL

n	shell name	capacity
1	k	2
2	l	8
3	m	18
4	n	32
5	o	50
6	p	72
7	q	98

NUMBER OF ELECTRONS IN A SUB-GROUP (SUB-SHELL)

l	sub-shell name	capacity
0	s	2
1	p	6
2	d	10
3	f	14
4	g	18
5	h	22
6	i	26

NUMBER OF ELECTRONS IN AN ORBITAL

All orbitals have a capacity of 2 electrons.

WAVE THEORY

The de Broglie wave velocity is

$$\text{v}_w = \lambda f = \frac{hf}{m\text{v}} = \frac{hf}{p}$$

The de Broglie wavelength depends on the mean electron radius and the principle quantum number.

$$\lambda = \frac{2\pi r}{n}$$

HEISENBERG'S UNCERTAINTY PRINCIPLE

$\epsilon_x \epsilon_p \sim h$
$\epsilon_t \epsilon_{\Delta E} \sim h$

where ϵ is the error or uncertainty in a quantity
x is position in meters
p is momentum ($m\text{v}$) in kg-m/sec
t is time in seconds
E is energy in joules
h is Planck's constant (6.626 EE–34 J-sec)

MASS-ENERGY EQUIVALENCE

$$E = mc^2$$

where E is in joules
m is in kilograms
c is the velocity of light (3 EE8 meter/sec)

BASIC PARTICLES

name	description and charge	symbol	C-12 scale mass (amu)
beta particle	electron (−1)	e or β	0.00055
neutron	neutron (0)	n	1.0087
proton	hydrogen nucleus (+1)	p	1.0073
alpha particle	helium nucleus (+2)	α	4.0015
deuteron	deuterium nucleus (+1)	D	2.0135
photon	gamma radiation (0)	γ	0.0000
neutrino	electron's neutrino (0)	ν_e	0
	muon's neutrino (0)	ν_μ	0
muon	mu-meson (−1)	$\mu-$	
pion	pi-meson (+1)	$\pi+$	0.15
	neutral pi-meson (0)	π_o	0.14
kaon	kappa meson (+1)	$K+$	0.53
	neutral kappa meson (0)	K_o	0.53

COMMON NUCLEAR CONVERSIONS

multiply	by	to obtain
amu	9.31 EE8	*eV*
amu	1.492 EE–10	*J*
amu	1.66 EE–27	kg
*Be*V	EE9	*eV*
erg	6.242 EE11	*eV*
erg	EE–7	*J*
erg	1.113 EE–24	kg
eV	1.074 EE–9	amu
eV	EE–9	*BeV*
eV	1.602 EE–19	*J*
eV	1.783 EE–36	kg
eV	EE–6	*MeV*
GeV	EE12	*eV*
grams	EE–3	kg
J	6.705 EE9	amu
J	6.242 EE18	*eV*
J	1.113 EE–17	kg
kg	6.025 EE26	amu
kg	5.610 EE35	*eV*
kg	8.987 EE16	*J*
MeV	EE6	*eV*

RADIOACTIVE DECAY

$m_t = m_o e^{-\lambda t}$

A_t = activity at time $t = A_o e^{-\lambda t} = \lambda N_t$

$$\lambda = \frac{0.693}{t_{\frac{1}{2}}}$$

$$\text{mean atom life expectancy} = \frac{1}{\lambda}$$

RELATIVITY

Let $k = \dfrac{1}{\sqrt{1 - \left(\frac{\text{v}}{\text{c}}\right)^2}}$

Then,

$$m_{\text{v}} = km_o \qquad \text{(mass increase)}$$

$$I_{\text{v}} = \frac{I_o}{k} \qquad \text{(length contraction)}$$

$$t_{\text{v}} = kt_o \qquad \text{(time dilation)}$$

$$f_{\text{v}} = f_o k(1 - \frac{\text{v}}{\text{c}}) \qquad \text{(frequency)}$$

CYCLOTRONS

$$r = \text{path radius} = \frac{m\text{v}}{Bq}$$

$$\omega = \text{angular velocity} = \frac{\text{v}}{r} = \frac{Bq}{m}$$

$$\text{v} = \text{linear velocity} = \frac{qBr}{m}$$

$$T = \text{time per revolution} = \frac{2\pi r}{\text{v}} = \frac{2\pi m}{qB} = \frac{2\pi}{\omega}$$

$$E = \text{kinetic energy} = \tfrac{1}{2}m\text{v}^2 = \tfrac{1}{2}m(r\omega)^2 = \frac{(qBr)^2}{2m}$$

$$\text{V} = \text{particle voltage} = \frac{\frac{1}{2}q(Br)^2}{m}$$

$$f = \text{required switching frequency} = \frac{1}{T} = \frac{qB}{2\pi m}$$

CONSISTENT UNITS FOR ABOVE EQUATIONS

r	meters
m	kilograms
v	meters/second
B	tesla (same as weber/meter^2) (multiply gauss by 0.0001 to obtain tesla)
q	coulombs
ω	radians/second
T	seconds
E	joules (same as kg-m/sec)
V	volts
f	hertz

LINEAR ACCELERATORS

The total energy possessed by a particle traveling at velocity v is the sum of its rest mass energy and its kinetic energy.

$$k = \frac{1}{\sqrt{1 - \left(\frac{\text{v}}{\text{c}}\right)^2}}$$

$$E_{\text{rest mass}} = m_o c^2$$

$$E_{\text{kinetic}} = m_o(k-1)c^2$$

$$E_{\text{total}} = m_o k c^2$$

$$m_{\text{v}} = km_o$$

$$\text{v} = c\sqrt{1 - \left(\frac{m_o}{m_{\text{v}}}\right)^2}$$

$$= c\sqrt{1 - \left(\frac{1}{k}\right)^2}$$

LIGHT AND ILLUMINATION

Wavelength, λ, may be measured in microns (EE–6 meters), millimicrons (EE–9 meters), or angstroms, Å, (EE–10 meters).

The *velocity of light* is approximately 3 EE8 m/sec, 9.84 EE8 ft/sec, 6.71 EE8 mph. Velocity, c, is constant although the frequency, f, will depend on the wavelength.

$$c = \lambda f$$

If the actual emitted frequency is f, and the relative separation velocity is v, the observed *Doppler* frequency is:

$$f' = \frac{f\left(1 - \frac{\mathrm{v}}{\mathrm{c}}\right)}{\sqrt{1 - \left(\frac{\mathrm{v}}{\mathrm{c}}\right)^2}}$$

The energy (in joules) of a *quantum of energy* will also depend on the frequency. In the equation below, h is *Planck's constant* which has a value of 6.626 EE–34 joule/sec.

$$\Delta E = hf$$

The illumination, E, (in ft-candles) on a surface a distance r from an omnidirectional source radiating F lumens is

$$E = \frac{F}{4\pi r^2}$$

OPTICS

Total reflection from a transparent medium will occur when the incident angle exceeds the *critical angle.*

$$\phi = \arcsin\left(\frac{1}{n}\right)$$

n is known as the *index of refraction.*

$$n = \frac{\text{speed of light in the first medium}}{\text{speed of light in the second medium}} = \frac{\sin\phi_i}{\sin\phi_r}$$

The *mirror equation* relates the distance from the mirror to the object (VA), image (VB), focus (VF), and center of curvature (VC).

$$\frac{1}{VA} + \frac{1}{VB} = \frac{1}{VF} = \frac{2}{VC} \quad *$$

The *magnification* produced by a spherical mirror is

$$M = -\frac{VB}{VA} \quad *$$

The *lens equation* relates the distance from the lens center to the object (OA), image (OB), and the focus (OF).

$$\frac{1}{OA} + \frac{1}{OB} = \frac{1}{OF} \quad **$$

The *magnification* produced by a spherical lens is

$$M = -\frac{OB}{OA} \quad **$$

The *lens makers' equation* requires knowledge of the radii of curvature. Values of r are negative for concave surfaces.

$$\frac{1}{OF} = (n-1)\left(\frac{1}{r_1} - \frac{1}{r_2}\right)$$

* *VF* and *VC* are negative for convex mirrors. Therefore, *VB* will also be negative for convex mirrors.

** *OA* is positive to the left of the lens, and *OB* is positive to the right of the lens. *OF* is negative for a diverging lens.

If OF is measured in meters, $\frac{1}{OF}$ is the *power of the lens* as measured in units of diopters.

Diffraction can be achieved with a single slit. The equation relating the slit size (d), wavelength (λ), and the angle to the mth dark band (out-of-phase cancellation) is

$$m\lambda = d(\sin\theta)$$

Diffraction can also be produced with a *diffraction grating* which has a slit spacing d.

$$m\lambda = d(\sin\theta)$$

WAVES AND SOUND

Velocity, frequency, and wavelength are related:

$$\mathrm{v} = f\lambda$$

Velocity of a longitudinal wave in a solid or liquid is

$$\mathrm{v}_{\text{long}} = \sqrt{\frac{g_c E}{\rho}}$$

Velocity of a longitudinal wave in a gas is

$$\mathrm{v}_{\text{long}} = \sqrt{k g_c R T}$$

Velocity of a transverse wave in a taught wire depends on the wire tension (F), length (L), and total mass (m_t).

$$\mathrm{v}_{\text{trans}} = \sqrt{\frac{FL}{m_t}} \quad \text{(consistent units)}$$

The frequency of the mth overtone (same as the $(m+1)$th harmonic) in a taught wire is

$$f = \frac{(m+1)\mathrm{v}_{\text{trans}}}{2L}$$

The fundamental frequency is the first harmonic, so $m = 0$.

HEAT TRANSFER

Conduction

Through a single layer: $q = \frac{kA\Delta T}{L}$

Through several layers: $q = \frac{A\Delta T}{\sum \frac{L_i}{K_i}}$

Convection

Through a film: $q = hA\Delta T$

Through layers with films: $q = \frac{A\Delta T}{\sum \frac{L_i}{K_i} + \sum \frac{1}{h_j}}$

Logarithmic mean temperature difference:

$$\text{LMTD} = \frac{\Delta T_1 - \Delta T_2}{\ln \frac{\Delta T_1}{\Delta T_2}}$$

Radiation

The energy radiated from a hot body at temperature T (in °R) and with emissivity ϵ is

$$q = \sigma\epsilon A T^4$$

σ is the Stefan-Boltzmann constant, equal to 0.1713 EE–8 BTU/hr-ft^2-°R^4.

$\epsilon = 1$ for blackbodies.

The net heat transfer between two bodies is

$$q = \sigma A_1 F_{1\text{-}2}(T_1^4 - T_2^4)$$

If object 1 is small and completely enclosed by object 2, then $F_{1\text{-}2} = \epsilon_1$.

NOTES

RESPONSE VARIABLES

A dependent variable that predicts the performance of a system is known as a *response variable*. Position, velocity, and acceleration vary with time and are the dependent response variables. Time is usually the independent variable.

NATURAL, FORCED, AND TOTAL RESPONSE

Natural response is induced when energy is applied to an engineering system and subsequently is removed. The system is left alone and is allowed to do what it would do naturally, without the application of further disturbing forces. If a system is acted upon by a force which repeats at regular intervals, the system will move in accordance with that force. This is known as *forced response*. The natural and forced responses are present simultaneously in forced systems. The sum of the two responses is known as the *total response*.

FORCING FUNCTIONS

An equation which describes the introduction of energy into the system as a function of time is known as a *forcing function*. The most common forcing functions used in the analysis of engineering systems are the sine and cosine functions.

$$F(t) = \sin \omega t$$

TRANSFER FUNCTIONS

The ratio of the system response (output) to the forcing function (input) is known as the *transfer function*, $T(t)$.

$$T(t) = \frac{R(t)}{F(t)}$$

Transfer functions generally are written in terms of the s variable. This is accomplished, if $T(t)$ is known, by taking the Laplace transform of the transfer function. The result is the *transform of the transfer function*, typically just called the *transfer function*.

$$T(s) = \mathcal{L}(T(t))$$

PREDICTING SYSTEM RESPONSE

Assuming that $T(s)$ and $\mathcal{L}(F(t))$ are known, the response function can be found by performing an inverse transformation.

$$R(t) = \mathcal{L}^{-1}(R(s)) = \mathcal{L}^{-1}(\mathcal{L}(F(t))T(s))$$

POLES AND ZEROS

A *pole* of the transfer function is a value of s that makes $T(s)$ infinite. Specifically, a pole is a value of s which makes the denominator of $T(s)$ zero. A *zero* of the transfer function makes the numerator of $T(s)$ zero. Poles and zeros can be real or complex quantities. Poles and zeros can be repeated within a given transfer function — they need not be unique.

A rectangular coordinate system based on the real-imaginary axes is known as an *s-plane*. If poles and zeros are plotted on the *s-plane*, the result is a *pole-zero diagram*. Poles are represented on the pole-zero diagram as ×'s. Zeros are represented as ○'s.

PREDICTING SYSTEM RESPONSE FROM POLE-ZERO DIAGRAMS

Poles on the pole-zero diagram can be used to predict the usual response of engineering systems. Zeros are not used.

pure oscillation: Sinusoidal oscillation will occur if a pole-pair falls on the imaginary axis. A pole with a value of $\pm ja$ will produce oscillation with a natural frequency of $\omega = a$ radians/sec.

exponential decay: Pure exponential decay is indicated when a pole falls on the real axis. A pole with a value of $-r$ will produce an exponential with time constant $\frac{1}{r}$.

damped oscillation: Decaying sinusoids result from pole-pairs in the second and third quadrants of the s-plane. A pole-pair having the value $r \pm ja$ will produce oscillation with natural frequency of

$$\omega_n = \sqrt{r^2 + a^2}$$

The closer the poles are to the real axis, the greater will be the damping effect. The closer the poles are to the imaginary axis, the greater will be the oscillatory effect.

STABILITY

A pole with a value of $-r$ on the real axis corresponds to an exponential response of e^{-rt}. Similarly, a pole with a value of $+r$ on the real axis corresponds to an exponential response of e^{rt}. However, e^{rt} increases without limit. For that reason, such a pole is said to be unstable.

Since any pole in the first and fourth quadrants of the s-plane will correspond to a positive exponential, a stable system must have poles limited to the left half of the s-plane (i.e., quadrants two and three).

FEEDBACK

The basic feedback system consists of a *dynamic unit*, a *feedback element*, a *pick-off point*, and a *summing point*. The summing point is assumed to perform positive addition unless a minus sign is present.

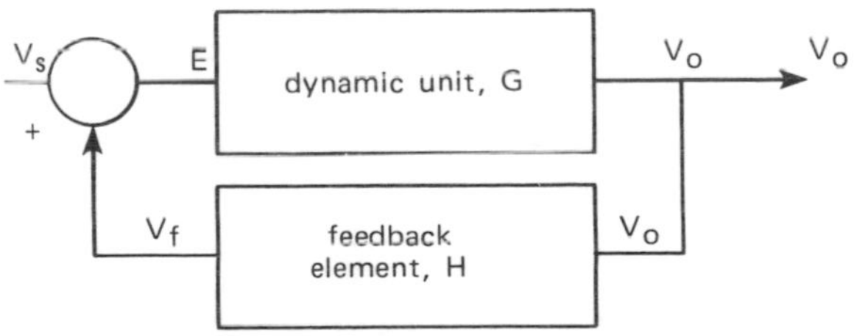

The dynamic unit transforms E into V_o according to the *forward transfer function, G*.

$$V_o = GE$$

For amplifiers, the forward transfer function is known as the *direct* or *forward gain*. The difference between the signal and the feedback is known as the *error*.

$$E = V_s + V_f$$

E/V_s is known as the *error ratio*. V_f/V_s is known as the *primary feedback ratio*.

The pick-off point transmits V_o back to the feedback element. The output of the dynamic unit is not reduced by the pick-off point. As the picked-off signal travels through the feedback loop, it is acted upon by the *feedback* or *reverse transfer function, H*.

The output of a feedback system is

$$V_o = GV_s + GHV_o$$

The *closed-loop transfer function* (also known as the *control ratio* or the *system function*) is the ratio of the output to the signal.

$$G_{\text{loop}} = \frac{V_o}{V_s} = \frac{G}{1 - GH}$$

NOTES

INDEX

Quick – I need additional study materials!

Please rush me the review materials I have checked. I understand any item may be returned for a full refund within 30 days. I have provided my bank card number as method of payment, and I authorize you to charge your current prices against my account.

	Solutions Manuals:
For the E-I-T Exam:	
[ ] Engineer-In-Training Review Manual	[ ]
[ ] Engineering Fundamentals Quick Reference Cards	
[ ] E-I-T Mini-Exams	
For the P.E. Exams:	
[ ] Civil Engineering Reference Manual	[ ]
[ ] Civil Engineering Sample Examination	
[ ] Civil Engineering Quick Reference Cards	
[ ] Seismic Design	
[ ] Timber Design	
[ ] Structural Engineering Practice Problem Manual	
[ ] Mechanical Engineering Review Manual	[ ]
[ ] Mechanical Engineering Quick Reference Cards	
[ ] Mechanical Engineering Sample Examination	
[ ] Electrical Engineering Reference Manual	[ ]
[ ] Chemical Engineering Reference Manual	[ ]
[ ] Chemical Engineering Practice Exam Set	
[ ] Land Surveyor Reference Manual	[ ]
Recommended for all Exams:	
[ ] Expanded Interest Tables	
[ ] Engineering Law, Design Liability, and Professional Ethics	

SHIP TO:

Name ______________________

Company ______________________

Street ______________________ Apt. No. ________

City ______________________ State ______ Zip ________

Daytime phone number ______________________

CHARGE TO (required for immediate processing):

______________________ ________

VISA/MC/AMEX account number expiration date

____

name on card

____

signature

Send more information

Please send me descriptions and prices of all available E-I-T and P.E. review books. I understand there will be no obligation.

____

____

____

____

A friend of mine is taking the exam too. Send additional literature to:

____

____

____

____

I disagree...

I think there is an error on page ________. Here is the way I think it should be.

Title of this book: ENGINEERING FUNDAMENTALS QUICK REFERENCE CARDS, Third Edition, third printing

[ ] Please tell me if I am correct.

Contributed by (optional):

____

____

____

Affix
Postage

PROFESSIONAL PUBLICATIONS, INC.
1250 Fifth Avenue
Belmont, CA 94002

Affix
Postage

PROFESSIONAL PUBLICATIONS, INC.
1250 Fifth Avenue
Belmont, CA 94002

Affix
Postage

PROFESSIONAL PUBLICATIONS, INC.
1250 Fifth Avenue
Belmont, CA 94002